Biobanks and Tissue Research

The International Library of Ethics, Law and Technology

VOLUME 8

For further volumes:
http://www.springer.com/series/7761

Christian Lenk · Judit Sándor · Bert Gordijn
Editors

Biobanks and Tissue Research

The Public, the Patient and the Regulation

Springer

Editors
Christian Lenk
Humboldtallee 36
37073 Göttingen
Germany
clenk@gwdg.de

Judit Sándor
Central European University
Nádor utca 9
1051 Budapest
Hungary
sandorj@ceu.hu

Bert Gordijn
Dublin City University
Residences Campus, Glasnevin
Dublin 9
Ireland
bert.gordijn@dcu.ie

ISSN 1875-0044 e-ISSN 1875-0036
ISBN 978-94-007-1672-8 e-ISBN 978-94-007-1673-5
DOI 10.1007/978-94-007-1673-5
Springer Dordrecht Heidelberg London New York

Library of Congress Control Number: 2011933663

Printed on acid-free paper

Springer is part of Springer Science+Business Media (www.springer.com)

Contents

Contributors

Petra Bárd Center for Ethics and Law in Biomedicine, Budapest, Hungary, bardp@ceu.hu

Katharina Beier Department of Ethics and History of Medicine, University of Göttingen, Göttingen, Germany, katharina.beier@medizin.uni-goettingen.de

Florence Bellivier University Paris Ouest Nanterre la Défense, Paris, France, florence.bellivier@wanadoo.fr

Alessandra Bernardi Istituto Oncologico Veneto-IRCCS, Via Gattamelata, 64-35128, Padova, Italy, alessandra.bernardi@ioveneto.it

Nikola Biller-Andorno Institute of Biomedical Ethics, University of Zurich, Zurich, Switzerland, biller-andorno@ethik.uzh.ch

Jasper Adriaan Bovenberg Legal Pathways Institute for Health and Biolaw, Aerdenhout, The Netherlands, jabovenberg@xs4all.nl

Emma Bullock Department of Philosophy, University of Birmingham, Birmingham, England, ECB767@bham.ac.uk

Asta Cekanauskaite Department of Medical History and Ethics, Vilnius University, Vilnius, Lithuania, asta.cekanauskaite@sam.lt

Vilius Dranseika Department of Medical History and Ethics, Vilnius University, Vilnius, Lithuania, vilius.dranseika@gmail.com

Bernice S. Elger Center for Legal Medicine, University of Geneva, Geneva, Switzerland, Elger@unige.ch

Eugenijus Gefenas Department of Medical History and Ethics, Vilnius University, Vilnius, Lithuania, eugenijus.gefenas@mf.vu.lt

Bert Gordijn Institute of Ethics, Dublin City University, Dublin, Ireland, bert.gordijn@dcu.ie

Nils Hoppe Institute for Philosophy, CELLS – Centre for Ethics and Law in the Life Sciences, Leibniz Universitaet, Hannover, Germany, nils.hoppe@agmb.uni-hannover.de

Julie Kent Department of Health and Applied Social Science, Faculty of Health and Life Sciences, University of the West of England, Bristol, UK, Julie.Kent@uwe.ac.uk

Christian Lenk Department of Ethics and History of Medicine, University of Göttingen, 37073 Göttingen, Germany, clenk@gwdg.de

Adam McAuley School of Law and Government, Dublin City University, Dublin, Ireland, adam.mcauley@dcu.ie

Federico Neresini PaSTIS Research Unit, Department of Sociology, University of Padua, Italy, federico.neresini@unipd.it

Christine Noiville University Paris 1 Panthéon – Sorbonne, UMR 8103, President of the CEES (Socio-Ethico-Economical Committee) of the HCB (High Council of Biotechnologies), Paris, France, noiville@univ-paris1.fr

Renzo Pegoraro Fondazione Lanza, Via Dante, 55-35139, Padova, Italy, renzo.pegoraro@fondazionelanza.it

Judit Sándor Central European University, 1051 Budapest, Hungary, sandorj@ceu.hu

Jurate Serepkaite Department of Medical History and Ethics, Vilnius University, Vilnius, Lithuania, jurate.serepkaite@gmail.com

Claudio Tamburrini Centre for Healthcare Ethics, Stockholm University, Stockholm, Sweden, claudio.tamburrini@philosophy.su.se

Ruud ter Meulen School of Social and Community Medicine, Centre for Ethics in Medicine, University of Bristol, Bristol, UK, r.terMeulen@bristol.ac.uk

Fabrizio Turoldo Fondazione Lanza, Via Dante, 55-35139, Padova, Italy, fabrizio.turoldo@unive.it

Heather Widdows Department of Philosophy, University of Birmingham, Birmingham, England, H.Widdows@bham.ac.uk

Elizabeth Yuko Institute of Ethics, Dublin City University, Dublin, Ireland, elizabethyuko@gmail.com

Part I
Biobanks, Tissue Research and the Public

Chapter 1
Introduction

Christian Lenk, Judit Sándor, and Bert Gordijn

1.1 Tissue Collections and Public Trust

The procurement and storage of human tissue and body parts has a long tradition in human history. The development of Western science – especially anatomy – was based, among other prerequisites, on the collection of human corpses and their parts. Similarly, a long-standing Western religious tradition involved retaining the remains of saints and holy objects for religious practices. Although the Enlightenment led to a kind of alienation of the public with regard to such habits and practices for religious purposes, our contemporary museums are full of human remains which were collected for scientific demonstration, curiosity, or both. For example, there was a discussion in Germany in 2010 as to whether two heads of Maori warriors could be shown in public (in an anthropological collection) or should be given back to the people of New Zealand. The *German Hygiene Museum* in Dresden has the complete and well-prepared body of a Lilliputian on exhibition, with unknown scientific value. And finally, thousands of spectators in Europe visited the *Body Worlds* exhibition of Gunther von Hagens. Beneath this open public fascination with the secrets of the human body in its various forms, institutions of medical therapy and research routinely store thousands of samples each day. In particular, the organ retention scandals in Ireland and the United Kingdom revealed that storing tissues and body parts was often done without the consent or even the knowledge of the concerned persons or their relatives in the past.

Although extensive tissue collections were built up in the past – for example in pathology institutes – this did not cause much ethical or legal discussion, most likely because of the diagnostic and therapeutic context. However, the situation obviously changed some years ago with the introduction of genetic analysis in tissue research and the new methodology of biobanks (i.e., the systematic and large-scale collection of body material for genetic analysis and linkage with personal health data)

C. Lenk (✉)
Department of Ethics and History of Medicine, University of Göttingen,
37073 Göttingen, Germany
e-mail: clenk@gwdg.de

C. Lenk et al. (eds.), *Biobanks and Tissue Research*, The International Library of Ethics, Law and Technology 8, DOI 10.1007/978-94-007-1673-5_1,

as well as genome-wide association studies. Tissue research as such is usually not invasive, when, for example, tissue which was already taken in the course of biopsies or other surgical interventions can be used for research purposes. Therefore, the existing risks of tissue research clearly do not fall under the category of physical harm. Rather, the crucial point seems to be the linking of private data on lifestyle with the outcomes of large-scale genetic studies; namely, the genetic testing of large groups of donors which are not necessarily patients. Such an expansion of genetic research and knowledge regarding genetic dispositions is obviously a new feature of research, while the social impact of those new methods is still unclear. The previous collection of "person-related" data was sparse, selective and isolated. This situation is changing at present in many areas such as the World Wide Web and social or economical activities, as well as in medical research.

Thus, a future scenario is emerging in which there is always existing data stored on every citizen in many areas of life (i.e. social life, social contacts, financial situations, professional backgrounds, living and working conditions, health, and also the genetic outfit) by public or private players. If a person or institution had the opportunity to combine and use this data altogether, fundamentally new dimensions of knowledge about a person or a larger group of people might well occur. The shift towards this potential scenario will therefore challenge our existing ideas of privacy and the normal situation of relative anonymity that most of us experience in public, and in turn, is likely to alter these ideas and perceptions dramatically. Astonishingly, the citizens themselves often reveal their data – step by step – to the public with some unintended side effects. Such behaviour is contradictory to known principles of data protection, such as the principle of "data austerity". The idea behind this principle is that data which will not be stored can also not be misused. Therefore, the default position should be that data should only be stored in cases with a definite and reasonable purpose which is accepted by the concerned person. However, today's techniques – such as automatic data storage in the context of internet usage or marketing practices – aim to circumvent this principle.

It is a fundamental principle of the EU Data Protection Directive that a person should have control over his or her personal data (cf. the Directive's Art. (12)). Some countries have had bad experiences with data collection (e.g. Germany) or lack of provisions made for data protection (e.g. the UK). Most likely resulting from the problematic experiences in Germany under totalitarian rule, there was an acrimonious debate about the accomplishment of a population census in the 1980s. It resulted in a landmark decision of the German Federal Constitutional Court[1] which was subsequently very influential for all matters of data protection in Germany. One newly-developed principle of data protection from this decision was a *right to informational self-determination*. From the ethical perspective, such a right can be interpreted as a part of a broader and more general right to self-determination, which today is usually seen as one of the most fundamental human rights.

[1]The so-called "Volkszählungsurteil": Urteil vom 15. Dezember 1983 (BVerfGE 65, 1). Available at: http://zensus2011.de/fileadmin/material/pdf/gesetze/volkszaehlungsurteil_1983.pdf (accessed 17 March 2011).

In 2007, it caused a government crisis in the UK, when two unencrypted CDs with 25 million personal records from the UK Department for Revenue and Customs (HMRC) got lost in the post. The CDs contained "the names, addresses and bank details of 9.5 million adults and the names, dates of birth and National Insurance numbers of all 15.5 million children in the country."[2] Due to such experiences, it is a central goal of genetic research with human tissue and biobanking to establish a system of data protection that is as safe as possible. This task is especially important in the case of vulnerable patient groups like children, HIV patients or schizophrenic patients who have been systematically involved in recent biobank projects. However, securing data protection in this kind of project is not an easy task because there are often scientific requirements stipulating that the data input and data output should be possible from a huge number of partner institutions (for example, general practitioners are responsible to feed data from their patients directly into a central database; or, international research groups can gain access to data of a national biobank). Therefore, data protection in the area of human tissue research is a very important, complex, difficult and sensitive issue.

Thus the question of trust comes into the discussion about tissue research and biobanks. Indeed, a considerable number of publications examine this issue. The individual must be able to trust public and other research institutions that information disclosed for scientific purposes will not be misused or abused for other purposes; for example, that scientific activities will not lead to single persons or groups of patients or donors being disadvantaged or socially stigmatised. These concerns were also explicitly considered in the *EU Convention on Human Rights and Biomedicine* and the *EU Data Protection Directive*.[3] However, a current study out of Germany shows that concerns over genetic discrimination are not unwarranted.[4] Therefore, it is obviously not only in the interest of patients and donors, but also of researchers and research institutions, that strict confidentiality in this area of research is ensured. From a socio-economical perspective, trust means that an interpersonal transaction can take place without further negotiations and work, and therefore without additional "transaction costs". Hence, trust is clearly a necessary precondition of projects which need a huge amount of such transactions, such as biobank projects. On the other hand, the distrust of the public in such projects would be an eliminating criterion for their further work. All parties therefore have an interest to find arrangements that will facilitate donor confidence in research.

---

[2] G. Rayner, Ch. Hope, and A. Porter, "Ministers 'ignored data security warnings'," *The Telegraph*, November 22, 2007, accessed October 21, 2010, http://www.Telegraph.co.uk.

[3] EU Convention on Human Rights and Biomedicine, 4. IV. 1997, Art. 11 – *Non-discrimination*: "Any form of discrimination against a person on grounds of his or her genetic heritage is prohibited." EU Data Protection Directive (95/46/EC), Art. 8.1: "Member States shall prohibit the processing of personal data revealing racial or ethnic origin, political opinions, religious or philosophical beliefs, trade-union membership, and the processing of data concerning health or sex life."

[4] Th. Lemke, "'A slap in the face': An exploratory study of genetic discrimination in Germany," *Genomics, Society and Policy* 5 (2): 2009, 22–39.

1.2 Tissue Research Between the Individual and the Public

It is a longstanding principle of Western thought and one of the tenets of Enlightenment philosophy that a person cannot be somebody's property. Indeed, when one looks into the publications of that time, authors like the German philosopher Immanuel Kant went even so far as to proclaim that nobody, including the concerned person himself or herself, has property in his or her body and therefore can also not dispose freely of his or her own body. For example, that people should not be allowed to commit suicide, to mutilate themselves or to sell parts of their own bodies. This radical "unavailability" (in German: "Unverfügbarkeit") of the human body (i.e., the idea that the body, like the person, should not be used as a mere means), even for the person himself or herself, was also a platform for the protection of the individual against the exploitation by others or demands of the society.

However, the development of the biosciences established a situation where, for example, the leftovers of surgical interventions are no longer worthless waste, but become a valuable raw material for the progress of medicine. Therefore, there has been an explicit demand for the individual to take part in research projects of public importance, when this poses only a minor burden on the individual, such as in the area of tissue research. In the end, the question is whether the government can successfully claim a kind of "public interest" in the individual body and its parts. For centuries, the dissection of human corpses for scientific or educational purposes was carried out on bodies recruited from the poor, welfare recipients and convicted criminals. To save costs and to acquire an additional benefit from these groups, it was traditionally assumed that those persons had forfeited certain civil rights and claims after their death, and that their corpses could be used for anatomical examination and demonstration without their consent. Totalitarian regimes showed in the past and perhaps also still in present – for example, in the People's Republic of China – that the dead bodies of certain individuals are seen as a disposable commodity used for the benefit of society.[5]

In a sense, the new possibilities of tissue and organ transplantation, as well as medical tissue research, seem to blur the established normative borders between the rights and duties of the individual and the rights and duties of the collective. A further example of this trend is the discussion of the use of medical biobanks for forensic aims: whether it is acceptable for the police to gain access to tissue collections which were established for medical research or disease prevention. This example also shows the significance of adopting a human-rights or a utilitarian perspective respectively when it comes to the evaluation and solution of such problems. While it is the central aim of the human-rights approach to protect the individual citizen against the powers of the state and society, the utilitarian approach insists, with

---

[5] S.E. Forsythe, "China's organ market: A tale of prisoners, tourists, and lies," *The New Atlantis*, Summer 2009, 121–24.

some justification, on the assertion of the public good. In this regard, how we perceive the individual body and the individual genome is also of some interest. Do we share the opinion that the individual body and genome belong to the person who then has the right to control these entities? Or do we see the human genome – and then also the individual genomes – as a "common heritage of mankind" as argued by the Ethics Committee of the Human Genome Organisation. This latter conception of the human genome may sound altruistic and generous – but does it not also pose risks for the individual who possibly has something in himself or herself which is not fully owned by him or her?

1.3 The Rights of Donors and Patients

We normally distinguish between legal and moral rights. Legal rights are typically defined in national law (so-called "positive law") while moral rights result from (sometimes internationally, sometimes nationally) shared values, normative traditions, ethical guidelines or so-called "soft-law". Nevertheless, due to the fact that only a minor part of human co-operation and communal life can be ruled by positive law, there remains a huge part of public life which is guided by our perception of moral rights. The treatment of the human body, although it is regulated in many respects (for example, the appropriate treatment of the human body after death in funeral law) obviously leaves a lot of space for culturally-different normative traditions. The treatment of bodily material after the separation from the human body – and, for example, the discussion on benefit-sharing in the context of genetic research and biobanking – is a striking example of this thesis. In this context of property, justice and body material, it seems to be more a question of equity or even bioequity[6] than of positive law to find an adequate balance between the interests of patients and donors on one hand, and researchers on the other.

Very similar problems occur concerning the question of control over bodily material once it is separated from the living body. Should donors have the right to control the use of their samples, even when the property rights in this sample have been explicitly transferred to a research institution? And why should we aim to control our donated tissue for science? There are indeed a number of initiatives to integrate the perspectives of patients and donors into research activities and research institutions (for example, in the form of the integration of patient representatives into institutional boards, or regular donor meetings and information events). However, the participation and information of patients and donors in the practice of research is in the end different from an explicit right to control the use of one's sample.

The decisive point in this regard seems to be the invention of genetic analysis and the dual character of human tissue: on the one hand it is tangible material, but

[6]To use a term which was coined by the German lawyer Nils Hoppe in his book entitled *Bioequity – Property and the Human Body*. Surrey, UK: Ashgate Publishing (2009).

on the other hand it can reveal a number of very important characteristics about the person from whom it stems. In the German-speaking countries (Austria, Germany, Switzerland), which have a distinct tradition of "personality law", this leads to the interpretation that because of these "informational" or "intellectual" characteristics of genetic data derived from human tissue, there is a right to control the use of human tissue which is beyond the mere assignment of property in human tissue. The different treatment of anonymised (no link between the donor and the sample) and identifiable or pseudonymised (the identity of the donor can be reconstructed via a code) samples in tissue research also show the importance of the "personal relation" of human tissue to a specific donor. When the identity of the donor of a specific sample cannot be reconstructed, most of the international documents demand fewer requirements for the use of that sample.

A further important area regarding patients' rights is the question of an appropriate informed consent in the area of tissue research and biobanking. In this context, it has also been shown that biobank research is different from more traditional medical research projects, especially when compared to controlled clinical trials. In particular, this concerns the fact that biobanks are primarily systematic collections of different bodily materials: not single, exactly-defined research projects for their own. Most of the large epidemiological – but also disease-related – public biobank projects offer the possibility for external researchers to gain access to the stored material for scientific reasons. But specifically, the type of research projects which will be carried out with this material is, in this case, unknown at the time when the tissue is collected. Therefore, some authors conclude (cf. the contribution of Bullock and Widdows in Chapter 8, this volume) that informed consent as such is not an appropriate ethical-legal instrument for the regulation of research with human tissue. Such a perspective is also partly supported by the Declaration of Helsinki which is equally applicable to "research on identifiable human material and data" (Declaration of Helsinki, part A, para. 1). The respective passage (part B, para. 25) of the Declaration reads as follows:

> For medical research using identifiable human material or data, physicians must *normally* seek consent for the collection, analysis, storage and/or reuse. There may be situations where consent would be impossible or impractical to obtain for such research or would pose a threat to the validity of the research. In such situations the research may be done only after consideration and approval of a research ethics committee. (Italics by the authors)

The Declaration of Helsinki here supports the option to refer to the local research ethics committee. One of the cases where it is "impossible" to obtain the donor's informed consent is the case of deceased persons. However, the named solution is incompatible with jurisdictions which explicitly demand the consent of the relatives in such a case. Additionally, it is not really clear why the rights of patients to be informed should ever "pose a threat to the validity" of research outcomes. The waiver of patient information is also contradictory to Art. 7, 10, and 12 of the EU Data Protection Directive. This short overview already demonstrates the difficulties which are connected with the regulation of this important issue.

1.4 Types and Definitions of Biobanks

The most important distinction regarding research biobanks is the distinction between population (or epidemiological) biobanks and disease-related biobanks. While population biobanks focus on a larger population (with individuals with a normal, i.e. mixed health condition), disease-related biobanks typically only collect samples from a specific group of patients. For example, most university hospitals have established tumour biobanks for research, and the scientific aims for these projects lie therefore in the area of cancer research. While disease-related biobanks focus on the molecular or genetic particularities of one disease or a group of connected diseases, population biobanks can in principle be used for research on all kind of diseases, provided that it contains an adequate amount of samples of bodily material to draw conclusions on the connection between a specific genotype and a disease. While disease-related biobanks have more similarities with classical collections from pathology and a "retrospective" methodology (the samples are extracted for diagnosis or in the course of the treatment), population-based biobanks are definitely a new approach for genetic research with a "prospective" methodology (samples are collected from persons with and without manifest diseases and the development of health conditions is then followed via regular reports).

Although the most specific European recommendation document in this area, the *Recommendation Rec (2006)4 of the Committee of Ministers to member states on research on biological materials of human origin,* would in principle be valid for both types of biobanks, Art. 17 only contains a definition for population biobanks:

1. A population biobank is a collection of biological materials that has the following characteristics:

    i. the collection has a population basis;
    ii. it is established, or has been converted, to supply biological materials or data derived therefrom for multiple future research projects;
    iii. it contains biological materials and associated personal data, which may include or be linked to genealogical, medical and lifestyle data and which may be regularly updated;
    iv. it receives and supplies materials in an organised manner.

However, a broader definition of biobanks which would also contain disease-related biobanks would be *a collection of samples of body material which is connected with genetic data and/or health data from patients or donors (in general: associated personal data).*

1.5 Pseudonymisation and Anonymization

Regarding the connection between a tissue's source (the donor), the sample and associated personal data,[7] the most important distinction seems to be between *anonymized* and *pseudonymised* samples – a distinction which, in the practice of research, often causes confusion. *Anonymized* means that there is no link obtained between the donor and his or her sample. Neither the researcher nor the biobank know or can re-identify the original donor of a body material. Some critics say that this case is only theoretical regarding the genetic characteristics of bodily material and that genetic material can in the end – with appropriate efforts – always be matched to its source. But although this may be true regarding the technical side of the question, this does not exclude the fact that there are sufficient ethical and legal safeguards to prevent such reidentification. *Pseudonymised* means that a code or a specifying characteristic links the donor's identity with a sample, but that the donor's identity is concealed from an external person (i.e. from a researcher of an external research group who wants to use a biobank's tissue collection) who is given the code but not the donor's name, address, birth date and so on. Normally, only the biobank management has access to the list of codes which can re-identify the single donors.

The pairs of words *anonymized – pseudonymised* and *non-identifiable – identifiable* are equivocal, although Art. 3 of the Council of Europe's *Recommendation Rec(2006)4* provides further differentiation:

> Biological materials referred to in Article 2 may be identifiable or non-identifiable:
>
> i. *Identifiable biological materials* are those biological materials which, alone or in combination with associated data, allow the identification of the persons concerned either directly or through the use of a code.
>
> In the latter case, the user of the biological materials may either:
>
> a. have access to the code: the materials are hereafter referred to as "coded materials"; or
> b. not have access to the code, which is under the control of a third party: the material are hereafter referred to as "linked anonymised materials".
>
> ii. *Non-identifiable biological materials*, hereafter referred to as "unlinked anonymised materials", are those biological materials which, alone or in combination with associated data, do not allow, with reasonable efforts, the identification of the persons concerned.

However, the distinction between (a) and (b) seems to have no major significance, either in scientific practice, or in ethical or legal theory. *Recommendation Rec(2006)4* also draws a normative conclusion from the distinction between

[7] *Personal data* is defined in the EU Data Protection Directive (95/46/EC) as "any information relating to an identified or identifiable natural person ('data subject'); an identifiable person is one who can be identified, directly or indirectly, in particular by reference to an identification number or to one or more factors specific to his physical, physiological, mental, economic, cultural or social identity; [. . .]".

identifiable or pseudonymised and non-identifiable and anonymised samples, i.e. that it is preferable in the context of the protection of privacy to use anonymised samples and that the use of pseudonymised samples has to be justified (Art. 8). However, most of the large biobank projects currently use pseudonymised samples, so that the work with anonymised samples is rather unusual.

1.6 Chapter Summaries[8]

Chapter 2 by *Julie Kent* and *Ruud ter Meulen* starts with the assumption that a primary purpose of regulation is to secure public trust. In the United Kingdom in the 1990s, public trust in those who procured, stored and used human tissues for research or other purposes was severely damaged following controversies surrounding the retention and use of post-mortem organs and tissue at the Bristol Royal Infirmary and the Alder Hey Royal Infirmary in Liverpool. Pathologists at both hospitals had retained large quantities of tissue and other remains of children without the knowledge or consent of their parents. Reports of unacceptable practices in Bristol and Alder Hey led to major revisions of laws and regulations of the use of all human tissue in the context of research and therapy, including the donation and transplantation of human organs. A new regulatory body, the Human Tissue Authority, was set up. Together with the previously-established Human Fertilisation and Embryology Authority, these two public bodies have had responsibility for key aspects of the regulation and oversight of human tissue use for research in the United Kingdom. The authors explore the role of these entities in securing public trust and confidence in current practices relating to research use of human tissue. These practices include the use of ova, embryos and aborted foetal tissue for stem cell research and most recently, the approved use of "human admixed embryos". Equally, the ethical principles underlying the current policies and legal frameworks in the United Kingdom regarding the research use of human tissue and their divergence from other European countries are discussed.

In Chapter 3, *Bernice Elger* and *Nikola Biller-Andorno* focus on the regulatory challenges of biobank research. From their perspective, biobank research is an essential element of new fields such as epigenetics, metagenetics and pharmacogenetics. Many European countries have invested considerably in this important research tool, which is expected to further the understanding of the interaction between genes and the environment and its implication for human diseases, as well as the development of efficient medical treatments. However, biobanks will be able to reach their full potential only if certain preconditions are met regarding the harmonisation of databank structures and their regulation. The collection and storage of DNA and cell tissue samples – as well as the collection of phenotypic, environmental and lifestyle data from medical records and patient questionnaires – needs to be standardised in order to guarantee sufficient quality of research, and to

[8]The summaries of the single chapters were provided by the authors.

permit collaboration. This process of technical standardisation is currently underway. Even more difficult is the harmonisation of ethical and legal frameworks. Over recent years, various national and international initiatives have been launched in this respect.

In Chapter 4, *Nils Hoppe* argues that the balancing exercises undertaken between individual interest and public interest in the context of human material procurement are distorted on the basis of unconvincing arguments. His chapter shows different bases for entitlements and makes a clear distinction between live and post-mortem procurement. The destination of the material is also argued to be of pivotal concern: where the material is used to save a life, where it is used to improve a person's health and wellbeing and where it is used for (potentially commercial) research – all of these scenarios demand different approaches to procurement governance. The text culminates in the proposal of an initial framework for a three-tiered system. Where the procurement is post-mortem and the material to be procured is necessary to save another's life, it is argued that there is no justification for withholding the material by means of an inter-vivos arrangement and it should be available without consent. Where the material is taken post-mortem and destined to improve another patient's health or wellbeing, the current system of free and voluntary donation can remain in place with all its limitations. Finally, where the material is taken from a live source and is required for research purposes, the source should be entitled to stipulate conditions (financial or otherwise) for the excision and further use.

Chapter 5 by *Federico Neresini* focuses on sociological considerations. The author argues that genetic biobanks are usually just considered a matter for scientists. Following this common point of view, we have, on the one hand, science, and on the other hand, society. On the contrary, he argues that biobanks could be considered an example of the mutual constitution of the scientific and the social. After a theoretical reframing of the relationship between science and society in the light of the Science and Technology Studies perspective, how biobanks perfectly embody this mixture of science and society will be clarified: they collect, purify and conserve organic material which is seemingly from an environment external to that of science, but to make it available for scientific research, they reorganise the environment according to its needs.

In Chapter 6, *Jasper A. Bovenberg* examines the relationship between law and economy in the field of tissue research. While the law allows universities and industry to capitalise on their contributions, it denies donors of biological materials both the right to compensation for and the right to control the use of their contributions.[9] To resolve this donor "cash and control inequity", or donor "gains and governance deficit", Bovenberg's chapter explores a novel solution: both universities and industry capitalise on their contributions by contributing these to a

[9] J.A. Bovenberg, "Whose tissue is it anyway?" *Nature Biotechnology* 37: August 2005, reprinted in *Property Rights in Blood, Genes and Data: Naturally Yours?* Brill/Martinus Nijhoff Academic Publishers, Leiden, Boston (www.brill.nl), MA, 2006.

corporation in exchange for shares in the corporation's capital. This form of capitalisation triggers an obvious, but hitherto unasked question: if inventors and investors can contribute in exchange for shares, then why can't donors? Issuing shares for samples may seem an awkward fit, but it is a proper way to give sample donors a say in both the gains and the governance of their samples. In view of the widespread use in the biopharmaceutical industry of stock option plans (for employees, suppliers, accountants, lawyers), a share issue (or a donor stock option plan) to those who contribute indispensable material, does not seem out of place. Thus, a "shares for sharing" model would resolve a set of ethical-legal claims in one stroke. An additional benefit of using a share-based approach is that it bestows rights (both to govern and to gain) that are commensurate to the value of the contribution. Rather than providing tissue donors with a blanket power to unilaterally dictate the terms of commercialisation without due regard to the contribution of other stakeholders, providing them with equity allows for fine-tuning and tailor-made rights that reflect the proportionate value of this contribution in relation to the contributions of the other contributors (the inventors and investors). In sum, redress the donor cash and control inequity, with equity.

The important topic of informed consent in biobank and tissue research is the theme of Chapter 7, by *Eugenijus Gefenas* et al. The requirement for informed, express and specific consent is one of the key principles of research ethics that evolved as a reaction to the atrocities of the Nazis' medical experimentations, as well as a response to the unethical human experimentation revealed during the post-World War II period. Such consent is thought to be the default position in clinical research and any softening of the requirement is usually perceived as an exception which requires justification. However, in some areas of human research, the requirement is more often both weakened in practice and criticised by the members of research community and ethicists. In this chapter we discuss circumstances under which research on human biological materials is, in fact, conducted without specific consent or re-consent of a donor. First, the authors explore the research use of materials that were collected for broadly-defined research purposes for which broad consent was initially secured. Second, the authors discuss the possibility to waive consent in research use of biological materials that were initially collected for non-research purposes without consent for research use. Third, three alternative regulatory regimes allowing turning residual biological materials into research collections during the collection procedure are addressed. These alternatives that justify the so-called research storage of biological materials collected for non-research purposes can be based on presumed consent, precautionary consent, and no consent.

From a different perspective, the issue of informed consent is also analysed by *Emma Bullock* and *Heather Widdows* in their Chapter 8. The acquisition of fully informed consent presents a central ethical problem for the procurement and storage of human tissue in biobanks. The tension lies between the apparent necessity of obtaining informed consent from potential research subjects, and the projected future use of the tissue. Specifically, under the doctrine of informed consent, medical researchers are required to inform their potential research subjects about the relevant risks and purposes of the proposed research (Declaration of Helsinki, 2008).

However, because human tissue – when stored in biobanks – can be put to multifarious uses, the information that medical researchers are expected to divulge to their subjects is epistemologically inaccessible. Biobank researchers are thus thought to be unable to obtain informed consent from their subjects, making the practice ethically suspicious. We propose that such suspicions of ethical failure should be reconsidered by presenting two possible solutions. Firstly, the authors argue that the epistemological difficulty might be partially solved by adopting the "waiver model" of informed consent. Secondly, Bullock and Widdows put forward an argument that individual consent can be supplemented by group ethical models. We thus conclude that while informed consent is problematic for biobank researchers, alternative ethical solutions are available.

Chapter 9 by *Claudio Tamburrini* focuses on the possible forensic use of biobanks. Privacy is increasingly becoming a more serious concern in the context of biobanking. For that reason, the anonymisation and pseudonymisation of samples containing donors' data have recently attracted much research. The concern with privacy is particularly evident in the area of forensic uses of biobank data. More concretely, opponents have argued that forensic data bases (i) discriminate against certain social groups, particularly when the data are kept even after the suspect has been dismissed from the investigation or acquitted in trial; (ii) lead to miscarriages of justice, as it is shown by some cases where innocents were found guilty because of errors committed in genetic data analysis; (iii) can be misused by governments to control citizens through information storage that might be used against them in the future; (iv) violate donors' privacy, particularly as genetic data banks imply that confidential information about donors' – as well as their relatives' – propensity to develop certain diseases is collected and put at researchers' – or State authorities' – disposal. Finally, it is also argued that all these problematic aspects of forensic biobanking, as they were expressed in the objections above, (v) can be conducive to the discredit of genetic biobanks in general, thus weakening people's willingness to contribute their samples to the repositories. Common to all these objections is however a remarkable lack of conceptual accuracy regarding both the nature and the content of the so-called right to privacy. Obviously, this has a direct bearing on how the other objections should be judged.

In Chapter 10, *Katharina Beier* and *Christian Lenk* argue that it is by now a well-investigated fact that human tissue research and biobanking is not regulated by a common legal framework in Europe so far. This chapter aims to step beyond this rather descriptive finding. By focussing on central issues of biobank research, the authors do not only highlight common trends and perspectives in the regulation of human tissue research across the countries of the European Union and Switzerland, but also identify the ethical and legal foundations for some of the persisting differences in this field. Their analysis bears on the premise that certain countries hold similar research traditions and are united by common ethical and legal pathways for regulating research. Based on their distinction of seven country groups and their respective regulatory frameworks, the authors finally draw some overall conclusions regarding the future regulation and potential legal harmonisation of this field within the European Union.

In Chapter 11, *Elizabeth Yuko, Adam McAuley* and *Bert Gordijn* begin their chapter by stating that a State's national interests determine that State's attitude towards international regulation. States' national interests differ – even States that share common features and characteristics. This is reflected in the approach of Ireland and the United Kingdom to the international regulation of research involving human tissue.[10] Ireland and the United Kingdom have a similar approach to many aspects of research involving human tissue, dictated by their membership of the European Union and Council of Europe. However, realism plays a crucial role in Ireland and the United Kingdom's regulation of research involving human embryos. Ireland's prohibitive approach is dictated by moral conservatism which can be traced to the significant influence of the Roman Catholic Church. The liberal approach of the United Kingdom reflects a society that seeks to accommodate science, business and religion.

Chapter 12 by *Renzo Pegoraro, Allesandra Bernardi* and *Fabrizio Turoldo* addresses the questions of procurement, storage and transfer of tissues and cells for non-clinical research purposes in the European Union's East-Mediterranean Countries (Cyprus, Greece, Italy, Malta and Slovenia). In particular, the authors describe the current local situation and explore the main similarities and differences between national legislation and ethical guidelines, and the application of European Union legislation in each country. Several ethical issues (e.g. the protection of confidentiality, and informed consent) have already found answers in the law, at least partially. Besides, some acts, even if they do not specifically pertain to biobanks for research, will be described in this text because many directives contained within them can be useful and are already applied in some biobanks for research purposes. The authors also explore in detail the national policies on sensitive data protection and donor's consent. Open issues are be addressed in the last part of the chapter, where the authors devote more attention to the ethical dimension of the topic. In the authors' opinion, decisive answers on some issues should not be given by the law because enforceable standards cannot be applied to ethical questions that require continuous debate and flexible solutions.

Chapter 13 by *Florence Bellivier* and *Christine Noivilles* focuses on the question of property in the human body. The only framework of rules applicable to the human body is that of objects or things, even though this unique and original thing must be considered specifically. Recognising the existence of practices that it sought to authorise under strict regulations (donations of human body parts and products, assisted reproductive technology, etc.), the French legislature set up a system of limited commodification. That is the reason why the concept of ownership to describe the relationship between the individual and its body may be appealing. Nevertheless, today it appears incomplete with regards to what probably constitutes the major issue in the analysis of the human body: conceptualising what happens,

---

[10] For the purposes of this chapter, residual embryos created following in vitro fertilisation (IVF) will be included alongside human tissue. The issue as to whether or not they are considered human tissue is debatable.

not at the beginning of the chain of the use of the parts and products of the human body, but following the changes and exchanges to which these human body parts and products give rise. In other words: how should access to biological resources be defined and regulated?

Chapter 14 by *Judit Sándor* and *Petra Bárd* provides a conceptual framework on the use and misuse of anonymity in biobank projects. The major point of departure is the comparative analysis of various functions of anonymity in the biomedical field. By applying this method, the authors also highlight differences between traditional and contemporary notions of anonymity. The other aim of the chapter is to distinguish between the overlapping terms of confidentiality, anonymity, data protection and privacy. While medical guidelines focus on the notion and technicalities of anonymization, legal provisions are based on the pillars of data protection norms, such as the possibility of identification, and the attachment to the original goal of data processing. The difference between these approaches resulted in various forms of ambiguities in the practice of biobanks. In addition to offering a theoretical framework, the authors analyse numerous models for regulation, with examples mainly from the Central and Eastern European region. The overview of the functions of anonymity reveals also to what extent anonymity can be regarded as an efficient solution for the regulatory challenges. The authors sketch out the most influential regulatory positions and analyse them critically, using a multidisciplinary approach.

Acknowledgement We would like to thank Silvia Schnorrer for her excellent work in the proofreading, indexing, and communicating with the contributors in the preparation of the final manuscript.

Chapter 2
Public Trust and Public Bodies: The Regulation of the Use of Human Tissue for Research in the United Kingdom

Julie Kent and Ruud ter Meulen

2.1 Introduction

In this chapter we will explore the extent to which the UK Human Tissue Authority (HTA) and the Human Fertilisation & Embryology Authority (HFEA) have successfully secured public trust and confidence in current practices relating to research use of human tissue. These practices include the use of ova, embryos and aborted fetal tissue for stem cell research and most recently the approved use of "human admixed embryos". We also reflect on whether the ethical principles underlying the current regulation have helped to secure public confidence in both the regulators and those who use human tissue for research purposes. Our discussion will help to clarify the ethical and social context of the current policies and regulatory frameworks in the UK regarding the research use of human tissue as well as their divergence from other European countries.

In April 2010 the UK Nuffield Council on Bioethics (NCB) launched a consultation exercise "to identify and consider the ethical, legal and social implications of *transactions* involving human bodies and *bodily material* in medical treatment and research" (p. 8). It was especially interested in considering the role of payments for such exchanges, consent, questions of ownership, the role of intermediaries and cultural and regulatory differences. The NCB is an independent body which focuses on ethical questions relating to biological and medical research, it has no statutory function or advisory role but does produce reports on its work which have the potential to influence UK policy. The launch of this latest consultation, 15 years after its previous report (Nuffield Council on Bioethics 1995), returns to continuing concerns about the ethical issues relating to the donation and use of human biological materials (HBM) and taking part in medical research.[1] While the NCB itself notes

[1] Although the Council is interested in making comparisons between donation of HBM and volunteering for first in human trials this goes beyond the scope of our discussion here.

J. Kent (✉)
Department of Health and Applied Social Science, Faculty of Health and Life Sciences, University of the West of England, Bristol, UK
e-mail: Julie.Kent@uwe.ac.uk

C. Lenk et al. (eds.), *Biobanks and Tissue Research*, The International Library of Ethics, Law and Technology 8, DOI 10.1007/978-94-007-1673-5_2,

the "considerable scientific, social and legal changes", which have taken place in the UK (and elsewhere), we see that in the last two decades questions and concerns about the exchange of HBM persist. It might seem surprising, in the wake of the significant overhaul of regulation since the early 1990s relating to HBM in the UK and across Europe, that further review is needed.[2] Yet there is much at stake here and many issues remain unsettled.

First, securing and maintaining public trust and confidence in medical treatment and research is frequently seen as essential if new therapies are to be developed (for example stem cell therapies), if existing practices are to be sustainable (for example, blood transfusion, organ transplantation, infertility treatment) and new knowledge is to be created. A key principle seen as securing and maintaining trust in medical research and treatment is the principle of informed consent: in the context of research with human tissue this means that a person has the right to be asked for consent for the removal, storage and research of any part of his or her body. This applies whether the tissue or body part comes from a living or deceased person. However, a continuing problem in consent procedures is whether consent should be *specific*, that is that the tissue may only be used for a clearly specified purpose, or that it can be *generic*, meaning that the sample may be used for various purposes, some of them going beyond the purposes mentioned in the information to the donor (for example because the research went in other directions). Second, while a new regulatory framework has emerged over this period, designed to secure that trust, and to regulate the collection and use of HBM for *therapeutic* and *research* use, evaluating whether it is "fit for purpose" is a legitimate and important ongoing activity. Third, from a sociological and anthropological perspective it is also of considerable interest to understand how, in practice, the regulations are socially shaped, interpreted, understood and *enacted* (Hoeyer 2009, 2010). A hotly debated issue is the extent to which increasingly we are witnessing the growth of a *market* for HBM and the emergence of diverse *tissue economies* and to what extent this transforms social relationships (Waldby and Mitchell 2006; Scheper-Hughes 2006; Dickenson 2007). While there has been a longstanding belief that donation or "gifting" of HBM may be linked to notions of citizenship and social solidarity – new technologies, potentialities and prospects, suggest that conventional understandings of these relationships may require revision and rethinking (Busby 2006, 2010b; Farrell 2006; Waldby 2006; Brown 2004; Faulkner et al., 2006). Part of the concerns about the emergence of tissue economies is the possibility that tissue donated for research and stored in public tissue banks (like UK Biobank) may be exploited by commercial enterprises, for example by the patenting of stem cell lines or gene sequences for diagnostic tests.

Internationally the European Group on Ethics in Science and New Technologies (EGE) has deliberated on ethical concerns relating to human tissue collection and

[2]We refer here to the EU Tissue and Cells Directive 2004, the Blood Safety Directives 2002/98/EC and 2004/33/EC, Regulation on Advanced Therapy Medicinal Products Regulation (EC) 1394/2007, and proposals for a new EU Directive on Organ Transplantation 2008/0238 (COD).

use in a number of its "opinions". For example in relation to the controversial use of human embryos in research it noted the lack of consensus about when "life" begins, the absence of agreement on embryo research within Europe and divergent national positions relating to whether or not such research should be permitted. The ethics of stem cell research has been the focus of a separate Opinion[3] which noted that whilst fundamental ethical principles could be identified, pluralism, cultural, religious and ethical diversity characterized the EU. Such diversity has been a focus of study for social and political scientists, ethicists and legal scholars (Salter and Salter 2007). The political difficulties which subsequently emerged around the funding of human embryonic stem cell research (hESC) in Europe under the 6th and 7th framework programme led, first to a moratorium on such research and then, in 2003, to a political decision to implement an ethics review process for projects which might be permitted under certain circumstances.[4] The process of *deriving* hESC was not to be supported by the framework research programme funding but the research use of the stem cells already derived became fundable (para III.4 Opinion 22). With respect to other types of human tissue use further ethical guidance has been given[5] though internationally there is continuing divergence between national legislation and practices relating to procurement and human tissue use in research. In regulatory terms, at EU level, the scope and remit of the 2004 Tissues and Cells Directive (TCD) was the collection and use of human tissues for *therapeutic* use. Tissue collection and use for non-therapeutic *research* was excluded. Regulation of such research was seen as falling within national jurisdictions. This allows for national diversity and in the UK means that regulation relates to the political histories of the regulatory institutions and specific features of the research culture. It is well recognised that national political culture contributes to policy formation and we argue it is therefore important to explain diversity and convergence within and between research and policy communities (Jasanoff 2005).

Additionally while our focus is on the procurement and *research* use of human tissue, in contrast to therapeutic use, evidence suggests that such distinctions may be less clear cut than might be expected. In the UK, a study of cord blood donation indicated that its potential value as therapeutic or research resource was entangled in the narratives constructed around donation to a public cord bank. "While the project information refers to the possibility of cord blood being used in research, the headlines of patient leaflets, press releases and the midwife's presentation all refer to the double 'gift of life' of a newborn who might save the life of a sick child" (Busby 2010a, 25). Furthermore "participants in the cord blood bank have to hold in the balance the possibilities of 'saving a life' for a patient who needs a transplant, and of the cells being used in the laboratory" (p. 25). This leads Busby

[3]Opinion No15 Ethical aspects of human stem cell research and use, 2000.

[4]Opinion No12 Ethical aspects of research involving the use of human embryo in the context of the 5th Framework Programme, 1998; Opinion 22 Recommendations on the ethical review of hESC FP7 projects, 2007.

[5]See http://ec.europa.eu/european_group_ethics/avis/index_en.htm (accessed 25 February 2011).

to conclude that the consent process may "oversimplify" complex issues relating to donation and its potential implications. So in the clinical setting, the distinction between therapeutic use of HBM and its use for *research* may be unclear (Parry 2006; Roberts and Throsby 2008). Moreover while consent has been seen as of fundamental importance for building and maintaining trust in the procurement of human tissue for research its operationalisation and meaning varies across settings as we shall elaborate below.

2.2 (Re) Building Public Trust in the UK

In Britain, in the 1990s the period between the Alder Hey and Bristol Royal Infirmary organs retention "scandal" and the passing of the UK Human Tissue Act 2004 (HT Act) may be characterized by a loss of public trust in what became seen as an outdated, patriarchal system of self regulation. It was found that there was a lack of oversight over the practices of pathologists and doctors and the retention of human tissues for diagnostic and research purposes (Bristol Royal Infirmary Inquiry 2000, 2001; Royal Liverpool Children's Inquiry 2001). During the inquiry into the children's heart surgery at the Bristol Royal Infirmary, it became apparent that it was common practice for pathologists to retain body material removed post mortem. This practice was wide spread among pathologists in England and Scotland. At the Alder Hey Hospital in Liverpool whole bodies of stillborn infants and whole organ systems were retained in the pathology department without parental knowledge. The subsequent review of organ and tissue retention practices in England and Scotland concluded that the regulations were lacking consistent and clear directions for post mortem examinations and retention of organs. To conduct a post mortem examination the hospital needed to comply with the wishes of the deceased or establish the deceased's lack of objection to procedures by questioning a relative. It was revealed that due to the vagueness of this procedure, hospitals assumed that the permission for a hospital autopsy could be taken as a full permission to retain the organs after the body was released for burial. The aim of the HT Act was to provide clarity in this context by way of a consistent legislative framework by replacing the principle of "lack of objection" by the principle of consent and by defining the purposes that require such consent (McLean et al. 2005). The source of power to retain, use and hold human tissue and organs resides in the consent either of the donor or otherwise of those who are qualified by an appropriate relationship with the deceased. The HT Act dealt with more than the removal and retention of bodily material for post mortem examination: the removal and retention of organs and tissue of the living for the purposes of research, audit, public health monitoring and education, but also the transplantation of organs and tissue for therapeutic purposes, became regulated under the new Act. The effects of the Act on the research (and clinical) community in the UK are still being felt. The Act makes the removing, storing or using of human tissue without consent, and also the taking and testing of DNA without consent, illegal.

The Human Tissue Authority (HTA) was established in 2005 with responsibility for regulating the collection and storage of "relevant material" – human tissue and cells from both the living and the dead for a wide range of purposes.[6] The collection and storage of tissue for *research* was just one area of its responsibility. It took 4 years, until 2009, for the HTA to produce a Code of Practice and guidance for those working in the research sector. This Code seeks to clarify the remit of the Authority stating that "the storage, as opposed to the use, of human tissue for research is licensed by the HTA. However, the consent requirements (of the HT Act) apply to its storage and use" (para 14). Crucially a distinction is made between the licensing requirements for *storage* and the approval of its use in research. Researchers must obtain approval for their research from a recognised Research Ethics Committee (REC) as part of the broader research governance framework in the UK discussed below. Consent for the material to be used in research should comply with the terms of the HT Act but anonymised samples used for research do not need consent if the project is approved by a REC.[7] Emphasising the importance of "proportionality", the HTA has sought to gain support for its activities through targeted efforts to engage "stakeholders" in the development and implementation of the new regulations.

In a survey of the research sector commissioned by the HTA (Opinion Leader 2009) evidence of the confusion surrounding regulation of human tissue research was reported. The HTA itself noted in its introduction to the report "participants found it difficult to distinguish between human tissue legislation and HTA regulation, and between the range of research activities beyond the HTA's remit, including ethics committee approval, funding and NHS Research and Development approval". Put another way, the role of HTA with regard to research use of human tissue was poorly understood by members of the research sector who took part in the survey. The authors wrote "there is confusion within the research community regarding the specific role and function of each of the bodies in charge of regulation and governance" (p. 8) and confusion about where researchers go for advice on human tissue research (p. 9). The regulatory requirements for research were perceived as complicated, difficult to understand and as having negative impacts on research, in particular making it harder to access samples, causing valuable samples to be lost, increasing time pressures and creating a burden of bureaucracy (only 9% of respondents thought legislation and regulation in this area had a positive impact on research p. 34). These findings confirmed those of a study of the use of fetal tissue in stem cell science discussed below. So how might we explain this confusion, complexity and these negative perceptions?

The HTA has a number of statutory functions arising both from the HT Act and the EU TCD. Its role is therefore wide ranging, encompassing diverse uses of tissues and body parts including for education, public display, transplantation and research. It has a licensing role for establishments wishing to store tissue for generic use, in

---

[6] See http://www.hta.gov.uk/ (accessed 25 February 2011).

[7] See HTA Code of Practice 1 Consent July 2006.

other words, what are sometimes referred to as tissue banks or biobanks require a licence if tissue is being *stored* for research purposes (140 licenses were in force at the time of writing). There are exceptions, for example, if the tissue is being stored en route to elsewhere or if the tissue is being used for a specific research project which has been approved by a REC (Human Tissue Authority 2009c). Yet the circumstances under which researchers require a licence to store tissue for research were seen as "complicated" (Opinion Leader 2009, 26). The survey suggested that public confidence in the processes by which tissue is obtained and used has been strengthened by the work of HTA, and researchers recognised the beneficial impacts on consent procedures. But the research community's own assessment of its negative effects could be seen as reflecting different political interests and professional concerns. Moreover while the HTA has a role with regard to the storage of research tissue, much human tissue research falls under the auspices of the wider research governance arrangements.[8]

2.3 Research Governance in the UK NHS

Concurrent with changes in human tissue regulations there have been considerable developments in wider research governance within the UK National Health Service (NHS). The 2001 NHS Governance Arrangements for Research Committees and subsequent Research Governance Framework (Department of Health 2001, 2005) marked a shift in thinking about research and signalled the establishment of a new governance structure (Eckstein 2007) which has been criticized for restricting access for research and needing to provide a clearer framework to assist the NHS to lead on clinical research (The Academy of Medical Sciences 2010). Furthermore, while RECs within the NHS were already established, in 1991 a Central Office for RECs (COREC) attempted to centralize administration of research applications. This was superseded by the National Research Ethics Service (NRES) in 2007.[9] There are now two types of NHS RECs: "recognised" and "authorised". Human tissue research may need approval from a "recognised REC" if it includes a clinical trial or an "authorised REC" may approve the research if it does not involve transplanting tissues or cells into patients but all NHS REC's are recognised for the purposes of the UK Human Tissue Act 2004.[10] So regulation of human tissue research is enacted through the work of both the HTA and REC's (and the Human Fertilisation and Embryology Authority as we will discuss below). Crucially the protection of public interests is framed in terms of protecting *donors* of HBM

[8] See 2008 NRES HTA Memorandum of Understanding, http://www.nres.npsa.nhs.uk/ (accessed 25 February 2011).

[9] See http://www.nres.npsa.nhs.uk/ (accessed 25 February 2011).

[10] There are approximately 100 NHS RECs for a fuller description see http://www.nres.npsa.nhs.uk/. Other RECs in Universities are not recognized as having powers to approve human tissue research.

through the procedures for ensuring that the principle of consent is upheld. However consent for research may be general or specific (see below) and the details of the research use are not necessarily specified to those living donors donating tissue or those in a qualifying relationship to the deceased and who give consent for tissue procurement (Human Tissue Authority 2006, para 105).

The deliberations of NHS RECs take place in secret and are not accessible to the wider public (Ashcroft and Pfeffer 2001). Membership of them comprises a minimum of a third of "lay person's" and experts from the health professions and science.[11] The authority and legitimacy of decisions by RECs is contested and complaints from the research community they regulate are common. Questions have also been raised about their accountability. It has been suggested that the ways in which they "do accountability" can be seen as undermining their "sociological legitimacy". While RECs tend to present their views as ethical "facts", the issues at hand are inherently opinions and judgements which are contestable (O'Reilly et al. 2009, 257). According to O'Reilly et al., there is a lack of evidence of "ethical reasoning" in the opinions of the RECs instead a number of tactics including – drawing on the authority of external bodies, referring to expert knowledge of the committee and blaming applicants for failures to comply with procedures; were used to account for their decisions. The claim to moral authority of RECs is based on their organisational and social location rather than any "appeal to the moral superiority of any ethical position" (Dixon-Woods et al., 2007, 800). A critical perspective on the research governance arrangements suggests that while the public may rely on REC's to protect their interests, the legitimacy of them, and the way they operate, is open to question and a focus for complaint from researchers.

2.4 Fetal Tissue – A Special Case

In a study which investigated the use of aborted fetal tissue in stem cell science, Pfeffer and Kent argued that in Britain the governance arrangements relating to fetal tissue use in research were confused, lacked transparency and were inconsistent with guidance on good practice relating to consent (Pfeffer and Kent 2006, 2007). They noted that little was known publicly about the use of aborted fetal tissue in research, there was no oversight of such use, no publicly available records were kept and amongst the research community practices varied in relation to the kinds of information given to women about the research for which the tissue would be used. Since the 1989 Polkinghorne Report which set out guidelines for the use of fetal tissue (Polkinghorne 1989) the "dead" fetus ex utero could be used for research and was regarded as like an organ donor by those who drew up the guidelines and others.

[11] Since the implementation of the Clinical Trials Regulation in 2004 the terms lay and expert are defined by NRES. See Information Paper on Membership of RECs v4.1 July 2009 at http://www.nres.npsa.nhs.uk/aboutus/about-recs/rec-membership/ (accessed 25 February 2011).

How death was defined or established was itself problematic since the usual criteria of brain stem death could not apply to the undeveloped fetus (Pfeffer and Kent 2007; Kent 2008). Women were framed in the guidelines as in need of protection from themselves and from researchers. The principle of "separation" underpinning the guidelines meant that the researcher should not have direct contact with the woman being asked to donate tissue; and the decision to abort the fetus should be separated from, and precede, the decision to donate tissue for research. Aborting a fetus in order to procure the tissue has been against the law since the 1967 Abortion Act but separation between the researcher and the woman became controversial when "The Royal College of Obstetricians and Gynaecologists claimed it is having an obstructive effect on research in fetal medicine, where "fresh" fetal material collected in utero is required. Furthermore, it objected to the inference that its members who undertake clinical research are incapable of conducting themselves ethically in relation to patients" (Pfeffer and Kent 2007, 436), (RCOG and Royal College of Pathologists 2001; RCOG 2004). Additionally while the Polkinghorne guidelines recommended that the method of termination and clinical care of the woman should not be influenced by the needs of researchers for the tissue, later Department of Health guidance permitted modification to the termination procedure "where it poses either the same or less risk to the women, has been approved by a REC, and been agreed to by the woman" (Pfeffer and Kent 2007; Department of Health 2002). Evidence of modifications was found where manual rather than mechanical extraction was used to increase the likelihood of the tissue being intact and therefore easier for researchers to dissect (Kent 2008).

There were important differences between the guidance on consent set out in the 1989 Polkinghorne Code and those statutory provisions enshrined in the 1990 Human Fertilisation and Embryology Act (HFE Act) which were designed to afford the in vitro embryo protection. These differences, including the emphasis placed on broad consent for use of fetal tissue and the much more specific and directed consent for use of embryos in research, reflected different views of women undergoing abortion and those having fertility treatment (Pfeffer and Kent 2007). The Polkinghorne guidelines were paternalistic and were later criticised for the insistence on broad consent, with the Department of Health proposing that specific consent for use in non-therapeutic research (including stem cell research) was more appropriate (Department of Health 2002). Following the HT Act, fetal tissue is now no longer distinguished from other living tissue "fetal tissue is subject to the same consent requirements under the HT Act as all other tissue from the living. However, because of the sensitivity attached to this subject, it is good practice to always obtain consent for the examination of fetal tissue and for its storage or use for all scheduled purposes" (Human Tissue Authority 2009a, para 157).[12] The HT Act and later HTA

[12]There is a further statutory consent exception for the use and storage of human tissue for research where all of the following criteria apply: tissue is from a living person; and the researcher is not in possession, and not likely to come into possession of information that identifies the person from whom it has come; and where the material is used for a specific research project approved by a recognised research ethics committee (Human Tissue Authority 2009c, para 41).

Code of Practice on Consent superseded, but did not fully review the Polkinghorne guidelines, as a number of aspects of these guidelines were seen as outside the remit of the HTA. "However, it should be noted that guidance within the Polkinghorne guidelines which recommended that in the context of giving consent, women should not know the purpose for which the fetus would be used, or whether it would be used at all, is now superseded by guidance within this code on valid consent, which must be based on the person's understanding of what the activity involves" (Human Tissue Authority 2009a, para 160).

"Valid consent" may be generic or specific and "generic consent typically only applies to research (see below). If conducting research on samples of tissue, it is good practice to request generic consent because this avoids the need to obtain further consent in the future" (Human Tissue Authority 2009a, para 36). Findings from Pfeffer and Kent's research showed that amongst the research community using fetal tissue, practices varied about the extent to which consent was general or specific and the extent of details of the research project which were given to women. This indicated both uncertainty about what was required, or considered good practice, and differing interpretations of the guidance and requirements of the RECs approving the research at that time. There were also variations in practice relating to disposal of unused tissue which in some cases complied with guidelines on "sensitive disposal" (Royal College of Nursing 2002) and in others followed usual practice for waste tissue disposal in laboratories.[13] Fetal tissue may be treated both as "waste" and as "a cadaver", its meaning and status is highly unstable across settings and in different social contexts (Kent 2008). What was also highlighted was the very different regulatory burden on researchers who use fetal tissue from those using embryos, something which researchers themselves saw as paradoxical and contradictory. Regulation of fetal tissue was seen as "light touch" compared to embryo research (Kent 2009).

2.5 From Embryos to "Human Admixed Embryos"

Regulatory discourse constructs different kinds of "regulatory objects" and embryos and fetuses may be seen as distinct "regulatory objects" which have been regulated separately (Kent 2009, 2012).[14] Since the 1980s the "embryo research debate" has been a focus of study for sociologists, ethicists, political scientists and legal scholars (Mulkay 1997). In the UK anti-abortionists mobilised around proposals put forward by Warnock to regulate, but permit embryo research. Other lobby groups date from the same period. In 1985 the PROGRESS Campaign for Research into Human Reproduction was launched "a pro-research coalition of patients, doctors, scientists and parliamentarians" which aimed to promote greater awareness of the

[13] See Human Tissue Authority 2009b, para 91–123 on Disposal of tissue following pregnancy loss.

[14] http://www.york.ac.uk/res/sci/events/FinalConfPres/Kent.pdf (accessed 25 February 2011).

benefits of scientific research and medical advancements that could be achieved.[15] Debates about embryo research have in the UK, as elsewhere, been tied to moral concerns about the destruction of embryos – those created as part of IVF treatment but also as a result of deliberate pregnancy termination. But while historically, organised opponents of embryo research were organisations formed in protest of abortion, interestingly the terms of the debate shifted. The regulatory framework for embryo research set up under the Human Fertilisation and Embryology Act in 1990 (HFE Act) creating the Human Fertilisation and Embryology Authority (HFEA), separated its function and remit from the regulation of abortion. The HFEA model meant that researchers wishing to use pre-implantation embryos in their research were, from 1990, legally required to seek approval from the authority and obtain a licence for their use. Details of research projects using embryos are now published and publicly available.[16] A key feature of policy making at that time was the level of public involvement (and protest) in the national debate. This has continued to be the case as the actions of the HFEA, have continued to attract public attention and criticism from some interest groups (Kent 2009, 2012).

> Arguments that scientific progress should be facilitated by a regulatory and legal framework won the debate in 1980s and characterised science policy in the UK from that period. But during the 1990s there was a growing awareness that public trust in scientists was in decline and a belief that it could be enhanced by better regulation and greater public engagement on scientific issues. The HFEA's job was both to secure public trust and facilitate the science and delivery of fertility services. Despite some criticisms and scrutiny of its role by House of Commons Science & Technology Committee in 2005 and 2007, proposals in the Draft Human Tissue and Embryos Bill (2007) to merge it with the HTA were strongly opposed by scientists and clinicians who lobbied for it to be retained. (Kent 2009)

The RCOG and PROGRESS opposed the merger because embryos should, it was argued, still to be regarded as a special case, as distinctive from other bodily tissues because of their reproductive potential and therefore deserving of special and separate protection. Moreover the HTA was considered an "unproven body" and merger with it would, it was argued, "dilute" the work of the HFEA (RCOG 2005, para 71). Kent suggests that controversy around the proposals in the Bill to merge the two regulatory bodies were intertwined with the arguments to support the use of "admixed human embryos" being put forward by scientists. Scientists in Britain, notably Stephen Minger and Lyle Armstrong, lobbied hard to secure support for the licensing of research which used animal eggs to create embryos for research. With the assistance of the Science Media Centre[17] they helped create what Minger termed "a new consensus between science and government".[18]

---

[15] http://www.progress.org.uk/ (accessed 25 February 2011).

[16] http://www.hfea.gov.uk/ (accessed 25 February 2011).

[17] http://www.sciencemediacentre.org/ (accessed 25 February 2011).

[18] Minger was speaking at a meeting of the ESRC Genomics and Research Forum Event held on 12 March 2009 in London – Conscience or Consultation? The HFE Act: a retrospective.

The creation of new biomedical entities through the mixing of animal and human cells has provoked vigorous public debate about the boundary between human and non-human species in the UK and in Europe (Brown and Michael 2004, Brown et al., 2006). Risks of contamination to humans and ethical concerns have been highlighted. The possibility of creating a "hybrid" or chimeric embryo using SCNT was regarded as especially problematic and was the subject of policy review in the UK between 2006 and 2008. Researchers argued that using animal eggs for SC research was more ethical than harvesting human eggs for SC research and that since these were to be used only as "research tools" and not for therapies there were no risks to human health. Controversially, licenses for the use of "animal human hybrids" were granted to the Newcastle and Kings College London research teams by the HFEA in advance of the review of law. The government made a "u-turn" on its policy between 2006 and 2007, first prohibiting it and then supporting the creation of human admixed embryos for research which was formally permitted in the new 2008 HFE Act under the jurisdiction of the HFEA (HFEA 2007a).

The British Government has been predominantly pro-science and supported embryo and stem cell research through what is frequently characterised as "a strict but permissive" regulatory approach (Pfeffer and Kent 2007). In contrast to the HTA, the HFEA has enjoyed the support of the clinical and research community which it regulates. One might say it has enjoyed greater "social legitimacy" despite criticisms from some other interest groups.[19] The HFEA's efforts to increase transparency and give greater public access to information about embryo (and stem cell) research has been directed towards securing greater public trust and confidence in its activities.[20] In addition to its statutory function enshrined in the 1990 Act and subsequent HFE Act 2008 relating to *research* it became, since 2006, a competent authority, responsible for implementing the 2004 EU TCD but specifically in relation to therapeutic use of gametes and embryos. While the HTA, as we have seen above, is responsible for all other human tissues. Unusually then in the UK there are two distinct and separate authorities responsible for the regulation of human tissues with a distinctive and separate structure and legislation relating to human embryos, gametes and the new biological entities now known as "human admixed embryos".[21] This has created a complex "regulatory maze" which is especially complex for stem cell researchers (Kent 2009, 2012).

---

[19]For example Comment on Reproductive Ethics (CORE) see http://corethics.org/ (accessed 25 February 2011).

[20]See press releases of HFEA for 7th May 2004, 25th November 2004 at http://www.hfea.gov.uk/ (accessed 25 February 2011) and cited in Pfeffer and Kent (2007).

[21]For a simple and excellent presentation on what these new entities are and how they are produced see http://www.wellcome.ac.uk/About-us/Policy/Spotlight-issues/Human-Fertilisation-and-Embryology-Act/Humanadmixedembryos/index.htm (accessed 25 February 2011).

2.6 Stem Cells – A New Regulatory Challenge

Another area of recent controversy in the UK, as elsewhere in Europe, has been the procurement of human ova for SC research. The HFEA's policy came under public scrutiny and was debated alongside a "consultation exercise" (Roberts and Throsby 2008, HFEA 2006, Devaney 2008). The HFEA did not sanction payment for eggs but decided that women would be allowed to donate their eggs to research, both as altruistic donors or in conjunction with their own IVF treatment. It argued that given that the medical risks for donating for research are no higher than for treatment, it is not for the HFEA to remove a woman's choice of how her donated eggs should be used (HFEA 2007b). Importantly women who were not undergoing IVF treatment would be permitted to make "altruistic" donations to research in addition to those who were already having treatment. Funded by the UK Medical Research Council, a Newcastle research team was the first in the UK to gain approval to pay towards the costs (approximately half) of infertility treatment if a woman agreed to donate half her eggs for research use.[22]

Justification for the scheme was framed first in terms of the need for larger numbers of eggs for SC research, second, for the *potential* therapeutic benefits which SC research might deliver in the longer term, thirdly as offering an opportunity, at no *extra* risk to young women for reduced costs, and by implication, increased access to IVF treatment. At the time of writing, the parallel study of the social and ethical issues related to such a scheme has not yet reported its findings.[23] Roberts and Throsby's (2008) analysis of representations of the proposed Newcastle scheme pointed out that rhetorical devices were used which elided the distinction between treatment and research, eggs and embryos and the act of donation and selling. They highlighted the burden placed on infertile women (not all women), who are already confronted with challenging experiences, to support SC research in such schemes. Unlike the donation of "spare" embryos in the case of egg sharing, the distinction between treatment and research is "not so easily maintained: donation to research is a *condition* of (subsidised) treatment and, while embryos can be donated *after* treatment, egg sharing depends on the relinquishing of eggs *during* treatment" (Roberts and Throsby 2008, 164). So, they argue, the Newcastle scheme overcame the illegal practice of egg selling through a process of aligning "sharing" for research with existing treatment practices (egg sharing for treatment) and normalising it in the context of other medical research.

The ethical issues arising in relation to women donating eggs who were not undergoing IVF treatment and who therefore were not already committed to hormonal stimulation for egg production provoked divided views. Some argued that the risks of egg collection were unjustifiable. Opposition to the use of human ova in research has centred both on the effects of increasing demand for human ova and

[22]See Press release September 2007 http://www.nesci.ac.uk (accessed 25 February 2011).

[23]This study was funded by the MRC at the same time and is directed by Prof Erica Haimes, see http://www.ncl.ac.uk/peals/research/project/2744 (accessed 25 February 2011).

creating a market for them and the health risks for women associated with "egg harvesting" and led to a controversial international campaign against the practice[24] (Beeson and Lippman 2006).

So the HFEA had an important role in facilitating access to eggs for stem cell research and, since 2001, the licensing of stem cell research which uses embryos.[25] It has frequently been upheld as a model of regulation which supports science and which it is claimed, is admired by other countries. Its role in relation to stem cell research is crucial to the enrolment of "donors" and procurement of embryos and eggs. However, once cells are derived from them the HFEA no longer has jurisdiction. Rather the cell lines, or new biological entities, created become redefined as regulatory objects which fall under the jurisdiction of the HTA and the Medicines and Healthcare products Regulatory Authority (MHRA) if they are to be used therapeutically. The transformation from a human embryo to hESC line, or from a human egg to a SCNT line, sets the researcher who has entered the regulatory maze along a different pathway. These pathways were most recently demonstrated in a new "route map" to guide researchers, later developed into a "tool kit".[26] Research has indicated that the boundaries between the three regulatory authorities with responsibilities for different aspects of SC research (and manufacture) have been contested and negotiated. The political processes which have sought to stabilise these institutional boundaries in the context of the distinct statutory roles of each agency (HFEA, HTA, MHRA), may be seen as directly associated with the emergence of these new bio-objects as part of the co-construction of socio-technical dimensions of "life" (Kent 2008, Brown et al., 2006). The ontological status of embryos and gametes which emphasised their reproductive potential in contrast to other human tissues, became embedded in a distinctive and separate institutional structure. Stem cell research created challenges to these arrangements which have been settled through conflict, compromise and contradiction. New mixtures of animal and human cells have been captured under HFEA jurisdiction but stem cell lines, as cultural products, fall outside its remit.

The HTA, as we have seen, had a different political history and responsibility for tissues of the living and dead, while MHRA is the competent authority with responsibility for the approval of new therapies.[27] The Gene Therapy Advisory Group (GTAC) has formally taken on responsibility for ethical review of clinical trials of

---

[24]Known as the "Hands off our ovaries" campaign. See http://handsoffourovaries.com/ (accessed 25 February 2011).

[25]In 2001, the Human Fertilisation and Embryology (Research Purposes) Regulations were enacted. These extended the purposes for which an embryo could be created.

[26]See Interim UK Regulatory Route Map for SC Research and Manufacture March 2009, and subsequent http://www.sc-toolkit.ac.uk/home.cfm (accessed 25 February 2011).

[27]A discussion of the approval of stem cell therapies goes beyond the scope of this chapter but relates to the 2007 EU Regulation of Advanced Therapy Medicinal Products REGULATION (EC) No 1394/2007. In May 2010 the European Medicines Agency noted that no stem cell therapies had been approved by them.

cell therapies *derived from stem cell lines* while other applications for clinical trials using human tissue and cells would normally be reviewed by RECs as described above.[28]

2.7 Generic and Specific Consent

Since the 2004 HT Act the legal regulation of human tissue research in the UK is based on the principle of consent: by defining the purposes of the specific research project and by asking for explicit consent it is assumed that individuals have control of the use of tissue and organs taken from their bodies. Practices like removal, retention, use and display of organs and tissue without consent have become illegal. The "ritual of informed consent" is seen as an effective procedure to protect individuals against unwanted use of parts taken out of their body (O'Neill 2003). However, there is a debate whether such protection is guaranteed in systems where the removal, retention, and storage of tissues is ruled by *generic* instead of *specific* consent. Tissue banks, like UK Biobank[29] are storing tissue on the basis of generic consent: future research using these tissue samples would become problematic if it required specific consent for each project. The ethical justification for generic consent in this context is based on the utilitarian logic that research with the data of these public banks will benefit the "public good" (Capps et al., 2008). Part of this logic is that the "public good" will be endangered when research projects are subjugated to specific consent procedures, in which not the public but the individual controls the resource (Capps et al., 2008). However, the utilitarian logic may lead to a general undermining of public trust and confidence in public authorities that govern the use of the tissue resources. The "tyranny of the majority" which is an inherent problem of utilitarianism might overrule the interests of the individual and could erode trust in governmental and non-governmental governance systems. A particular problem is the use of public banking resources for industrial research projects, which may conflict with the interests of the community. When such conflicts of interests become badly managed and private interests become dominant, public trust will decline. It remains to be seen to what extent the governance body of the UK Biobank and other governance bodies will be able to balance the various interests in the use of publicly stored human tissue.

---

[28]"GTAC oversees clinical trials involving cells derived from stem cell lines. A stem cell line is a permanently established culture of unspecialised cells derived from a single parental cell, or group of parental cells, that can (1) proliferate in vitro for a prolonged period when given appropriate nutrition and space and (2) be made to differentiate in culture into more specialised types of cells when given appropriate chemical or molecular cues. This includes cell therapies derived from: genetically modified cells; embryonic stem cell lines; multipotent stem cell lines; mesenchymal stem cell lines; foetal stem cell lines; induced pluripotent stem (iPS) cell lines". http://www.dh.gov.uk/ab/GTAC/Stemcelltherapy/index.htm (accessed 25 February 2011).

[29]See http://www.ukbiobank.ac.uk/ (accessed 25 February 2011).

While the principle of *generic* consent may open the door to unwanted use of data, the principle of *specific* consent raises problems as well. This principle is dominant in the NRES system which regulates and controls the use of human tissue in research projects in the NHS, though generic consent may be allowed if the data are fully anonymised as described above, However, specific consent, as sought by the RECs is not invariably better than generic consent (O'Neill 2003) and can raise new issues of accountability: "Complex forms that request consent to numerous, highly specific propositions may be reassuring for administrators (they protect against litigation), and may have their place in recruiting research subjects: yet they will backfire if patients or practitioners come to see requesting and giving consent as a matter of ticking boxes" (O'Neill 2003, 6). O'Neill makes a plea for genuine consent, that is consent where individuals can control the amount of information they receive and where they have a right to withdraw consent in case they disagree with the use of their tissue. Conditional consent, as advocated by National Ethics Committees in France and Germany (Dickenson 2006), is not practised in the UK. In fact, many researchers in the UK regard the donation of tissue a contractually binding transfer, with no further rights of the donor to withdraw the donation or to the future use (and benefits) of the tissue (Dickenson 2006).

2.8 Conclusions

We have shown that diverse uses of human tissue in research have provoked extensive public debate and policy changes over the past 30 years in the UK. Ethical concerns relating to human tissue research provoked very different responses in the Polkinghorne and Warnock reports 30 years ago. Embryo research has been facilitated through the regulatory framework established in 1990 and reaffirmed in the recent 2008 HFE Act. Despite objections from the anti-abortion lobby and some well organised interest groups a permissive pro-science consensus has prevailed and sustained support for the work of the HFEA (Jasanoff 2005). Support for embryo research has been maintained since the 1980s and regulated separately and more strictly than other human tissues. The HFEA's role has been to secure public trust in embryo research and most recently, in a review of its role, support from the scientific and clinical community was evident.

In the 1990s there was public outrage at the retention of organs, including children's body parts and fetal tissue, without consent and for research uses which were poorly managed. Since 2004 human tissue research became subject to new regulatory constraints in the wake of this. This lead to the establishment of the HTA whose role is poorly understood and it has not so far secured support from the research community or public trust.

The new biology produces new life forms and biological entities such as "admixed human embryos" and stem cell lines which has in turn created new challenges for regulators and led to new governance arrangements. In the UK the statutory bodies of the HFEA, HTA, MHRA, together with RECs and GTAC have responsibility for laboratory research and clinical trial governance. While their purpose is to protect the public interest they facilitate the procurement and research use of the tissues, cells and bodies of the public. However what has developed over time is a complex regulatory maze which has been criticised by members of the research community.

Since we began writing this paper recent events mean there are further twists in this tale. Political priorities change and in a recent review of "arms length bodies" by the new UK Government, and in the context of economic pressures to make financial savings from the public purse, in July 2010 it was announced that both the HFEA and HTA will be closed down (Department of Health 2010). Proposals to "rationalise the regulatory landscape", streamline procedures and reduce bureaucracy include recommendations that the functions of the two authorities are reallocated to other bodies and that a new "research regulator" be established pending the outcome of a review of the regulation and governance of medical research by the Academy of Medical Sciences. The Academy, whose primary concern is to promote medical science, wants to enhance the facilitation and translation of medical research into clinical practice and to promote the UK as a site for clinical trials. It recommended in an earlier report that improvements could be made to the regulation of medical research (The Academy of Medical Sciences 2010). Its review of research governance arrangements will inform the implementation of these latest plans to set up a new "research regulator" which will also supersede the current National Research Ethics Service. Crucially the regulation of research, it is proposed, should be separated from the other activities (e.g. licensing) of the HTA and HFEA which in turn are expected to be reallocated to other retained regulatory bodies. This raises questions about the implications for public trust in human tissue research. As we have seen the complex web of regulation surrounding human tissue research stems from the need to build and secure that trust but has also been shaped by strong political interests. Clinical and scientific interests have been important but so too have wider public interests. How the next phase of these debates plays out in light of these proposals remains to be seen and will require new legislation, but it seems unlikely that the old politics will be swept away by the economic pressures to come up with new organisational solutions to longstanding ethical, social and legal problems.

Acknowledgments The authors wish to thank the UK Economic and Social Research Council for the research funded by them which contributed to this chapter (Research Grant award RES-350-27-004).

References

Ashcroft, R., and N. Pfeffer. 2001. "Ethics Behind Closed Doors: Do Research Ethics Committees Need Secrecy?" *British Medical Journal* 322 (7297): 1294–96.

Beeson, D., and A. Lippman. 2006. "Egg Harvesting for Stem Cell Research: Medical Risks and Ethical Problems." Accessed February 25, 2011. http://www.rbmonline.com edn

Bristol Royal Infirmary Inquiry. 2000. *Interim Report, Removal and Retention of Human Material: Annex and Annex B*. London: Central Office of Information.

Bristol Royal Infirmary Inquiry. 2001. *Learning from Bristol; the Report of the Public Inquiry into Children's Heart Surgery at the Bristol Royal Infirmary 1984–1995*. London: HSMO.

Brown, N., A. Faulkner, J. Kent, and M. Michael. 2006. "Regulating Hybrids: 'Making a Mess' and 'Cleaning Up' in Tissue Engineering and Transpecies Transplantations." *Social Theory and Health* 4: 1–24.

Brown, N., and M. Michael. 2004. "Risky Creatures: Institutional Species Boundary Change in Biotechnology Regulation." *Health, Risk and Society* 6 (3): 207–22.

Busby, H. 2006. "Biobanks, Bioethics and Concepts of Donated Blood in the UK." *Sociology of Health & Illness* 28 (6): 850–65.

Busby, H. 2010a. "The Meanings of Consent to the Donation of Cord Blood Stem Cells: Perspectives from an Interview-Based Study of a Public Cord Blood Bank in England." *Clinical Ethics* 5: 22–27.

Busby, H. 2010b. "Trust, Nostalgia and Narrative Accounts of Blood Banking in England in the Twenty First Century Health." *An Interdisciplinary Journal for the Social Study of Health, Illness and Medicine* 14 (4): 364–382.

Capps, B., A. Campbell, and R. ter Meulen. 2008. *Access to the UK Biobank Resources: Concepts of the Public Interest and the Public Good*. UK: Biobank. Accessed February 25, 2011. http://www.egcukbiobank.org.uk/meetingsandreports/

Department of Health. 2001. *Governance Arrangements for Research Ethics Committees*. London: Department of Health.

Department of Health. 2002. *Human Bodies, Human Choices: The Law on Human Organs and Tissue in England and Wales – A Consultation Report*. London: Department of Health.

Department of Health. 2005. *NHS Research Governance Framework*. London: Department of Health.

Department of Health. 2010. *Liberating the NHS: Report of the Arms-Length Bodies Review*. London.

Devaney, S. 2008. "Breaches in Good Regulatory Practice – The HFEA Policy on Compensated Egg Sharing for Stem Cell Research." *Clinical Ethics* 3 (1): 20–24.

Dickenson, D. 2006. "The Lady Vanishes: What's Missing Form the Stem Cell Debate." *Bioethical Inquiry* 3: 43–54.

Dickenson, D. 2007. *Property in the Body: Feminist Perspectives*. Cambridge: Cambridge University Press.

Dixon-Woods, M., E. Angell, R. Ashcroft, and A. Bryman. 2007. "Written Work: The Social Functions of Research Ethics Committee Letters." *Social Science and Medicine* 65 (4): 792–802.

Eckstein, S. 2007. "Research Ethics Committees." In *Good Clinical, Laboratory and Manufacturing Practices Techniques for the QA Professional*, edited by N. Dent and P. Carson, 99–108. Cambridge, UK: The Royal Society of Chemistry.

Farrell, A.. 2006. "Is the Gift Still Good? Examining the Politics and Regulation of Blood Safety in the European Union." *Medical Law Review* 14 (2): 155–79.

Faulkner, A., J. Kent, I. Geesink, and D. Fitzpatrick. 2006. "Purity and the Dangers of Regenerative Medicine: Regulatory Innovation of Human Tissue-Engineered Technology." *Social Science & Medicine* 63: 2277–88.

Hoeyer, K. 2009. "Tradable Body Parts? How Bone and Recycled Prosthetic Devices Acquire a Price Without Forming a 'Market'." *BioSocieties* 4: 239–56.

Hoeyer, K.. 2010. "An Anthropological Analysis of European Union (EU) Health Governance as Biopolitics: The Case of the EU Tissues and Cells Directive." *Social Science & Medicine* 70 (12): 1867–73.

Human Fertilisation and Embryology Authority. 2006. *Donating Eggs for Research: Safeguarding Donors*. London: HFEA.

Human Fertilisation and Embryology Authority. 2007a. "Hybrids and Chimeras: A Report on the Findings of a Consultation." Accessed February 25, 2011. http://www.hfea.gov.uk/docs/Hybrids_Report.pdf

Human Fertilisation and Embryology Authority. 2007b. *Statement on Donating Eggs for Research*. London: HFEA, February 21.

Human Tissue Authority. 2006. *Code of Practice 1: Consent*. London: Human Tissue Authority.
Human Tissue Authority. 2009a. *Code of Practice 1: Consent*. London: Human Tissue Authority.
Human Tissue Authority. 2009b. *Code of Practice 5: Disposal of Human Tissue*. London: Human Tissue Authority.
Human Tissue Authority. 2009c. *Code of Practice 9: Research*. London: Human Tissue Authority.
Jasanoff, S. 2005. *Designs on Nature: Science and Democracy in Europe and the United States*. Oxford: Princeton University Press.
Kent, J. 2008. "The Fetal Tissue Economy: From the Abortion Clinic to the Stem Cell Laboratory." *Social Science & Medicine* 67: 1747.
Kent, J. 2009. "Enabling Stem Cell Science in the UK? A Short History of a Regulatory Maze, Beyond Pattison." Stem Cell Initiative Final Conference, Wellcome Trust, London, May 2009.
Kent, J. 2012. *Regenerating Bodies: Tissue and Cell Therapies in the 21st Century*. Abingdon: Routledge.
Mclean, S., A. Campbell, H. Harper, and K. Gutridge. 2005. *Scoping Study: Ethical and Practical Concerns Regarding Changes to Human Tissues Legislation*. Bristol: University of Bristol; Glasgow: University of Glasgow.
Mulkay, M. 1997. *The Embryo Research Debate Science and Politics of Reproduction*. Cambridge: Cambridge University Press.
Nuffield Council on Bioethics. 1995. "Human Tissue Ethical and Legal Issues." Available at: http://www.nuffieldbioethics.org/sites/default/files/Human%20tissue (accessed 14 March 2011).
Opinion Leader. 2009. *Impact of Legislation and Human Tissue Authority Regulation on Research*. London: Human Tissue Authority.
O'neill, O. 2003. "Some Limits of Informed Consent." *Journal of Medical Ethics* 29: 4–7.
O'reilly, M., M. Dixon-Woods, E. Angell, R. Ashcroft, and A. Bryman. 2009. "Doing Accountability: A Discourse Analysis of Research Ethics Committees' Letters." *Sociology of Health & Illness* 31 (2): 246–61.
Parry, S.. 2006. "(Re)constructing Embryos in Stem Cell Research: Exploring the Meaning of Embryos for People Involved in Fertility Treatments." *Social Science and Medicine* 62 (10): 2349–59.
Pfeffer, N., and J. Kent. 2006. "Consent to the Use of Aborted Fetuses in Stem Cell Research and Therapies." *Clinical Ethics* 1 (4): 216–18.
Pfeffer, N., and J. Kent. 2007. "Framing Women, Framing the Fetus: how Britain Regulates the Collection and Use of Aborted Fetuses in Stem Cell Research and Therapies." *BioSocieties* 2 (4): 429–47.
Polkinghorne, J. 1989. *Review of the Guidance on the Research Use of Fetuses and Fetal Material. Cm 762*. London: HMSO.
Roberts, C., and K. Throsby. 2008. "Paid to Share: IVF Patients, Eggs and Stem Cell Research." *Social Science and Medicine* 66 (1): 159–69.
Royal College of Nursing. 2002. *Sensitive Disposal of Fetal Remains*. London: Royal College of Nursing.
Royal College of Obstetricians and Gynaecologists. 2004. *Response to MRC Draft Code of Practice for the UK Stem Cell Bank*. London: RCOG.
Royal College of Obstetricians and Gynaecologists. 2005. *RCOG Ethics Committee Response to the Department of Health Review of the Human Fertilisation and Embryology Act*. London: RCOG.
Royal College of Obstetricians and Gynaecologists and Royal College of Pathologists. 2001. *Fetal and Perinatal Pathology Report of a Joint Working Party*. London: RCOG.
Royal Liverpool Children's Inquiry. 2001. *Report Ordered by the House of Commons*. London: The Stationary Office.
Salter, B., and C. Salter. 2007. "Bioethics and the Global Moral Economy: The Cultural Politics of Human Embryonic Stem Cell Science." *Science, Technology & Human Values* 32 (5): 554–81.

Scheper-Hughes, N. 2006. "Alistair Cooke's Bones: A Morality tale." *Anthropology Today* 22 (6): 10–15.

The Academy of Medical Sciences. 2010. *Reaping the Rewards: A Vision for UK Medical Science*. London: Academy of Medical Sciences. http://www.acmedsci.ac.uk

Waldby, C. 2006. "Umbilical Cord Blood: from Social Gift to Venture Capital." *Biosocieties* 1: 55–70.

Waldby, C., and R. Mitchell. 2006. *Tissue Economies: Blood, Organs and Cell Lines in Late Capitalism*. Durham: Duke University Press.

Chapter 3
Biobanks and Research: Scientific Potential and Regulatory Challenge

Bernice S. Elger and Nikola Biller-Andorno

3.1 Biobanks and Research: The Scientific Potential

In 2000, the completion of the draft human genome sequence was announced (Butler 2010). During the past 10 years the technical possibilities of automated data analysis of DNA samples and their bioinformatic processing have continuously and dramatically evolved. There has been considerable criticism of the "hype" around the sequencing. This is due to its focus on a race between rivalling scientific institutions and its overemphasis on the relevance of knowing the sequence of the "whole" human genome, fostering a public misunderstanding that "it's all in the genes", that simple gene defects could serve as a model for the most common diseases and that quick cures were virtually around the corner. Still, the sequencing of the human genome can be considered a milestone towards what has been termed the "GWAS era" (Latourelle et al. 2009; O'Brien 2009): human biospecimens, DNA, genotype, and clinical data are combined in so-called biobanks to carry out genome wide association studies (GWAS). They explore the interaction between genes and the environment as well as the implications for human diseases and medical therapies. The rising demand for human tissue in research illustrates the rapid expansion of the field (Womack and Gray 2009).

In Europe as well as globally, these collections of specimens, also called biobanks or genetic databases,[1] represent a significant amount of public investment and have become an important research tool comprising studies in new fields such as epigenetics (Kavikondala et al. 2010; Talens et al. 2010), systems

[1] In this chapter, the terms "biobank" and "genetic database" are used interchangeably to signify a collection of human biological samples that can be used for genetic analysis, including those that combine such samples with the results of genetic analyses and health or other data about the persons from whom the samples were collected. The category encompasses pathology collections, repositories for specific diseases (e.g. cancer registries), and population databases created to permit longitudinal studies of any disease or condition (see Elger et al. 2008, p. 1 note 1).

B.S. Elger (✉)
Center for Legal Medicine, University of Geneva, Geneva, Switzerland
e-mail: Elger@unige.ch

C. Lenk et al. (eds.), *Biobanks and Tissue Research*, The International Library of Ethics, Law and Technology 8, DOI 10.1007/978-94-007-1673-5_3,

biology (Diez et al. 2010), toxigenomics (Chung et al. 2009) and translational and basic science stem cell research (Bardelli 2010), including research on somatic-cell nuclear transfer (Jones and MacKellar 2009).

The European Science Foundation published a report in May 2009 which acknowledges the scientific importance of biobanks. The slogan "good biobanks for better health" (Reed and Bjugn 2010) characterizes the underlying public health objective. By combining data about environmental exposure with health outcomes and genetic analysis, epidemiological research, as well as research concerning specific diseases, can be advanced significantly.

3.1.1 Disease Types

Most cohort studies on different diseases nowadays have their own attached biobank (Garcia-Merino et al. 2010; Jiang et al. 2009). Biobanks are increasingly considered an indispensible tool in the search for answers to many health related questions, including public health concerns, as the titles of recent studies suggest:

> Do evolving practices improve survival in operated lung cancer patients? A biobank may answer. (Vlastos et al. 2009)
>
> Is smoking an independent risk factor for invasive cervical cancer? A nested case-control study within Nordic biobanks. (Kapeu et al. 2009)
>
> Overcoming the global crisis ... also for TB ... ? – yes, we can [if we use biobanks]. (Ottenhoff 2009)

Ottenhoff reminds the reader that tuberculosis (TB) causes almost two million deaths every year and argues that "high-quality clinical trial capacity and biobanks for TB biomarker identification" are important tools. They are increasingly used by public health researchers as well as WHO surveillance centres. Biobanks have become a "must" for research on many infectious diseases, as is illustrated by the opening of the King's College London (KCL) Infectious Diseases BioBank in 2007, which collects peripheral venous blood from patients infected with various pathogens including human immunodeficiency virus (HIV) (Williams, Mant, and Cason 2009). Another field in which biobank research contributes to an important public health goal is the use of blood products for transfusion purposes. The Blood and Organ Transmissible Infectious Agents (BOTIA) project has their own biobank in which paired donor-recipient samples are stocked for research (Lefrere and Coudurier 2009).

Human biobanks are of great scientific value to researching diseases where gene-environment interactions are complex. Examples are cardiovascular disease (Posch et al. 2009), neurological diseases (Teunissen et al. 2009), and especially most cancer types (Clement, Chene, and Degos 2009; Riegman, de Jong, and Llombart-Bosch 2010). Indeed, repositories of DNA, RNA, and serum samples play a key part in the investigation of the underlying causes of cancer development, progression, and prognosis. They are indispensable resources for the investigation of biomarkers which serve to detect cancers early and to predict treatment response (Ennis et al. 2009).

The past years have also shown a tremendous increase in the establishment of paediatric biobanks, beginning in oncology, but extending their scope recently to all sorts of paediatric diseases (Ebner et al. 2010). This rise has been triggered, among other things, by the fact that research with children is highly regulated since children are a vulnerable population. Biobank research has the advantage of being considered in most cases as minimal risk research, since there is no direct harm to children if their samples are examined (Garcia-Merino et al. 2010; Gurwitz et al. 2009).

3.1.2 Pharmacogenomics

Although disease-related biobanks were among the initial biobanks to have been established, media attention was first significantly raised in the context of population biobanks that announced "personalized medicine" as their main goal, as shown by the title of this journal article:

> With your genes? Take one of these, three times a day. (Abbott 2003)

The aim of pharmacogenetics is to lead to personalised therapy based on genetic profiling. Biobanks therefore have a noticeable place in drug discovery research. In the great majority of clinical trials drug companies submit to the FDA for approval, and provisions are made to sample and store blood for future genetic analyses (Abbott 2003). Pharmacogenetic and pharmacodiagnostic tools are used to improve drug efficacy and safety margins. For several years, interest has been centred on the genetic polymorphism of drug-metabolising enzymes such as cytochrome P450s (CYPs) and N-acetyltransferases (NATs), which have been studied in Caucasian, Asian and African populations (Gurwitz et al. 2005; Gurwitz and Pirmohamed 2010; Matimba et al. 2009). The pharmaceutical industry has expressed interest in using population biobanks to develop new targeted medicines. The pharmaceutical company Hoffmann-la Roche is said to have paid $200 million in order to obtain the rights to develop and market drugs resulting from genes that deCode had hoped to find for a dozen disorders through research involving the Icelandic national biobank (Durham and Hall 1999; Enserink 1998a, b; Lemonick 2006; Nutley 2002; Schwartz 1999).

3.1.3 National Biobanks

Many common diseases, such as cardiovascular and psychiatric conditions, are influenced by multiple genes. In order to determine the influence of groups of SNPs (single nucleotide polymorphisms, i.e. a form of DNA sequence variations) on drug responses in diseases, a large number of samples is required, and SNPs need to be searched across the entire genome (Abbott 2003). Modern high-throughput testing has enabled the establishment of research using large national biobanks at less cost. Iceland, Estonia, Denmark, Spain, and Croatia are examples of countries that have established their own national biobank. More and more countries are following their example (Andorno 2006; Kaiser 2002; Modin et al. 2010; Rudan et al. 2009; Zika et al. 2010). National biobanks are often advertised as the creation of "biovalue" (Mitchell 2010): Advocates call them an economic "resource" of interest not only to basic researchers and academic biologists, but also to pharmaceutical genomics companies that invest in diagnostic and clinical products. Yet, many large European DNA biobanks have encountered difficulties. The promises regarding their scientific or medical benefits were not fulfilled as quickly as researchers and industry had announced (Rose 2006).

3.1.4 The Importance of National and International Collaboration

One of the most critical factors in biobank research is the availability of a sufficient number of samples in order to ensure adequate powering of studies. Suitable sample sizes often cannot be obtained in single-center studies (Teunissen et al. 2009). Large national biobanks have been established to carry out research mostly on common diseases. In contrast to these national biobanks, sample collections dedicated to research on specific, less common diseases tend to be small and attached to a single university or hospital. Even in large national biobanks, the frequency of certain diseases is too low to justify specific studies. Networking between different biobanks, albeit still rare (Zika et al. 2010), therefore becomes more and more critical to remedy these shortcomings (Asslaber and Zatloukal 2007; Clement et al. 2009; Salvaterra et al. 2008; Yuille et al. 2009).

National and international collaborations between biobanks can only be efficient if a certain number of conditions are fulfilled. In order to carry out meaningful comparisons between samples and data, phenotypic information needs to be detailed and well standardized (Gurwitz and Pirmohamed 2010; Ritchie et al. 2010). The way in which samples are obtained and processed, including the time that elapses between the taking of samples and conservation measures, such as freezing, are of immense influence on the quality of samples, and the results of certain studies might vary simply because of different preparation procedures (Botling et al. 2009; Cardoso et al. 2010; Johnsen et al. 2009; Rudloff et al. 2010). Another important organisational aspect of collaboration and networking is the communication about availability of samples. A common way to do this is the establishment of catalogues that are widely accessible via the Internet, providing information on which institutions hold which types of samples (Chabannon et al. 2010). This requires, however, that institutions agree to collaborate and have policies as well as material transfer agreements that ensure comparable ethical and legal standards. Collaborations are hampered significantly if regulations of ethical and legal issues vary between different countries or even between institutions within the same jurisdiction.

It is therefore not surprising that major European funding agencies, such as the European Science Foundation, acknowledge not only the scientific importance of biobanks, but also the need for harmonization of databank structures and their regulation (Ballantyne 2008; European Science Foundation 2008).

3.2 Harmonization of Technical Procedures for the Preparation, Handling and Storage of Samples and Data

While biobanks have enormous and almost ubiquitous scientific potential in various areas of medicine, their impact and efficiency is significantly decreased if collections remain fragmented. The collection and storage of DNA, cell tissue samples, as well as the collection of phenotypic, environmental and lifestyle data from medical records and patient questionnaires need to be standardized in order to achieve sufficient quality of research and to permit collaboration within biobank networks.

International and European organizations of scientists are working on the standardization of technical procedures with varied success. Many technical aspects of biobanking, as well as the influence of epigenetics and metagenetics, concern both human and non human biobanks. International initiatives are therefore aiming to ensure global harmonized standards that overcome the traditional borders between human subject (human biobanks) and non-human subject (non-human biobanks) research (Day and Stacey 2008; ISBER 2009).

In the US, the National Cancer Institute (NCI 2007) has put considerable work into the elaboration of technical SOPs (standard operating procedures). In Europe, the Biological and Biomolecular Research Infrastructure (BBMRI) Program has convened a Pathology Expert Group Meeting that produced its own recommendations. These emphasize the role of pathologists (Bevilacqua et al. 2010). Worldwide, pathologists are handling the bulk of available specimens. They also act as gatekeepers to essential information which permits the identification of specimens.

> [Pathologists] make decisions on what should be biobanked, making sure that the timing of all operations is consistent with both the requirements of clinical diagnosis and the optimal preservation of biological products. (Bevilacqua et al. 2010)

Appropriate training of pathologists in institutions that are hosting biobanks is crucial in order to ensure not only that "the timing of all operations is consistent with both the requirements of clinical diagnosis and the optimal preservation of biological products" (Bevilacqua et al. 2010), but also to harmonize standard operating procedures (SOPs) and to fulfill international standards.

Among the technical aspects, the linkage of the biobank with existing databases, a hospital database or, often, local or national cancer registries is an overlooked aspect of the standardization of biobank-based studies. Up to now, the linkage of biobank material to cancer registry data as a way to enhance research protocols has rarely been examined and included in published recommendations (Langseth et al. 2010).

3.3 Harmonization of Ethical and Legal Issues Concerning Biobanks

The linkage between samples and data, especially if the latter are obtained from registries that contain personal identifiers, raises important ethical and legal questions regarding consent, privacy and management of information (Netzer and Biller-Andorno 2004). The harmonization of ethical and legal frameworks regarding biobanks has proven to be particularly thorny (Elger et al. 2008; Elger 2010). While the need for international guidelines has been widely recognized (O'Brien 2009), the existing regulatory framework remains a complicated patchwork of more or less contradictory local guidelines and laws. The Council of Europe's recommendation on research with biological material is a promising step in the right direction that has taken many years of preparation. The Organisation for Economic Co-operation and Development (OECD) published its own guidelines in autumn 2009, which will hopefully help to catalyze future harmonization of domestic laws. Indeed, following the example of Iceland, Estonia and Sweden, other European countries have issued laws during the past year (Spain) or have prepared law projects (Switzerland).

3.3.1 Biobanks and Classical Health Research Ethics

The development of international guidelines is taking time because biobank research is a challenge to classical health research ethics (Elger 2010; Elger and Caplan 2006). If fundamental principles such as informed consent and a strict definition of personal data are applied, biobank research becomes largely unfeasible or at least disproportionately costly. There is an ongoing dispute as to whether biobank research requires a redefinition of the balance between patients' rights and science/society's quest for efficient and affordable beneficial research, or whether problems can simply be resolved through an adequate interpretation of research ethics principles when applied to biobanks.[2] The latter approach means that the balance itself will not change, including the high value given to individual human rights concerning privacy, individual choices and control over body parts, and tissues. According to the former approach a value shift is necessary towards a greater weighting of community values, such as solidarity and altruism of tissue donors, and a restriction to individual autonomy based rights.

The debate might also be framed as including a question about paternalism: should individuals who participate in biobank research be allowed to wave future rights to control the use of their samples and data when they provide broad consent to future research studies, though they have not been informed about yet unknown details of these projects? Traditionally, research ethics contains a paternalistic element: the decision whether a human research study is considered too dangerous

[2]Indeed, using existing data and samples for secondary research purposes does not imply direct physical risks and could therefore justify broad, less informed consent.

to be acceptable is taken by a research ethics committee (REC). Research participants are not allowed to take risks if the REC considers them disproportionate to the benefits (Belmont Report 1979). Those who argue that the overall balance of individual rights and the interests of science/society should remain unchanged, claim that biobank research implies mostly minimal risk (Caulfield and Weijer 2009; Gurwitz et al. 2009; National [NBAC] 1999, pp. v, vi and 7). Allowing biobank participants to wave their individual right to truly informed consent for future studies involving their samples and data would therefore be acceptable without questioning the importance of classical informed consent in traditional clinical trials.

When it comes to discussing the ethical problems in their concrete contexts, the debate about whether fundamental values or their balance are changed or only adapted becomes less predominant. New guidelines show that advocates from both sides may interpret changes as being in line with their own framework and agree about the proposed measures. In the following part we will sketch out the recently proposed compromises and solutions for the three most important controversial issues: informed consent, privacy and returning of research results to participants.

3.3.2 Recent Developments Concerning Controversial Ethical Issues

Informed consent remains a controversial issue. If biobank research is evaluated within the framework of classical research ethics, it does not seem acceptable to allow research participants to consent to future studies with having received sufficient information. However, in recent years some evolution towards the acceptance of broad (partially uninformed) consent occurred in international guidelines. In 2008, the Declaration of Helsinki was revised, including a paragraph that softens consent requirements concerning research with identifiable human material and data:

> 25. For medical research using identifiable human material or data, physicians must normally seek consent for the collection, analysis, storage and/or reuse. There may be situations where consent would be impossible or impractical to obtain for such research or would pose a threat to the validity of the research. In such situations the research may be done only after consideration and approval of a research ethics committee. (World Medical Association [WMA] 2008)

Overall the ethical discussion has moved away from the almost exclusive concentration on one-time original consent towards the management of future new uses of collected material that had not been anticipated at the time of original consent. The OECD guidelines on human biobanks and genetic research databases (HBGRD) admit alternatives to traditional informed consent when they discuss the four major problems (see the bullet points in paragraph 3.1 of the OECD guidelines below) with which biobanks may be confronted if the original consent did not precisely cover future research uses of biological material and data.

3.1 Review processes, in accordance with applicable law, including research ethics committees or comparable oversight mechanisms, should be in place for use in cases where human biological materials or data are to be used in a manner not anticipated in the original informed consent process, including:

- for previously collected human biological materials or data where the use might deviate from the original consent;
- for cases where informed consent may not have been obtained at the time of collection;
- for determining when to seek re-consent;
- for use of human biological materials or data where consent was obtained using a broader or layered format for uses unspecified at the time of collection, especially in the case of large-scale genetic epidemiology studies (OECD 2009)

The OECD guidelines propose three different solutions to deal with future yet unknown projects involving human biological material and data from biobanks. Paragraph 4.5 presents the first two: new consent or a waiver of consent.

4.5 Where subsequent use of human biological materials or data is envisaged that would not be consistent with the original informed consent, a new consent should be obtained from the participant or from the appropriate substitute decision-maker, or a waiver of consent should be obtained from a research ethics committee or an appropriate authority, in accordance with applicable law and ethical principles pertaining to the protection of human subjects. (OECD 2009)

While new consent and waivers have been tools permitted in traditional research ethics, the acceptance of broad consent, according to the following paragraph of the OECD, is a step towards an adaptation of classical informed consent with respect to biobanks.

4.6 Where authorized by applicable law and the appropriate authorities, the operators of the HBGRD could consider obtaining a consent that will permit human biological specimens and/or data to be used to address unforeseen research questions. Participants should be fully informed of the breadth of such consent and there should be additional safeguards in place to ensure that participants are protected. (OECD 2009)

With this statement the OECD goes a step further than the Council of Europe, which does not use the term broad or general consent, although it contains a description of consent that could be interpreted as compatible with the broader type of consent.

10.2 Information and consent or authorisation to obtain such materials should be as specific as possible with regard to any foreseen research uses and the choices available in that respect. (COE 2006)

However, it should be noted that the OECD guidelines permit broad consent only if "additional safeguards" are in place. This is again an example of the regulations' shift of attention towards ongoing monitoring and management, away from a one time consent when participants enter a biobank study (Meslin 2010). The way in which ongoing control and oversight mechanisms could be standardized remains at present vague. A *cantus firmus* of the debate seems to be the fact that oversight mechanisms should be independent from funders and researchers (Secko et al. 2009).

The OECD frames their support for broad consent very cautiously by narrowing its use to jurisdictions where such practice is "authorised by applicable law and the appropriate authorities". In many European (?) countries, the legal framework concerning informed consent does not at present accomodate any broader standards. The same holds true for the United States. Since 2004, the Office for Human Research Protection (OHRP 2004) has somewhat circumvented the consent issue and enacted adaptations instead in the domain of *privacy* (Elger and Caplan 2006). It broadened the definition of non-identifiable (non-personal) samples and data. Coded material and data are considered non-identifiable if researchers or other users do not have access to the code. This permits researchers to use samples and data for further projects without the need of renewed consent or new approval of a REC as long as they use coded material in the aforementioned way.

In Europe, the position on the definition of personal data has not changed in the same way as in the US, although some evolution took place. A data protection working party of the European Commission proposed the following definition of anonymous data:

> Article 29 [...] "Anonymous data" in the sense of the Directive can be defined as any information relating to a natural person where the person cannot be identified, whether by the data controller or by any other person, taking account of all the means likely reasonably to be used either by the controller or by any other person to identify that individual. (Data Protection Working Party 2007)

The interesting development is that data could be considered anonymous eventhough it could still be possible to identify individual persons. However, protocols and procedures are in place that exclude this from happening, for example through technical means such as "cryptographic, irreversible hashing". Article 29 might, however, also be read in the sense that protocols and procedures are in line with a reversible coding of samples and data where it is a contractual arrangement that guarantees that researchers and users of the data do not have access to identifying information.

> Article 29 [...] In other areas of research or of the same project, re-identification of the data subject may have been excluded in the design of protocols and procedure, for instance because there is no therapeutical aspects [sic!] involved. For technical or other reasons, there may still be a way to find out to what persons correspond what clinical data, but the identification is not supposed or expected to take place under any circumstance. (Data Protection Working Party 2007)

It is noteworthy that the Working Party examined only the case of data. Whether the possible expansion of the definition of anonymous data may be extended to biological samples remains questionable. Therefore, biobanks in Europe are at present not allowed to circumvent consent and REC approval in the same way as is possible for government-funded research under OHRP provisions in the US to do so.

Besides some evolution concerning the issues of consent and privacy, the question of whether biobanks and/or researchers should *return research results to participants* has moved somewhat forward (Bovenberg et al. 2009). While in the past researchers were free to decide whether to communicate individual research results that fulfilled the requirements of validity, significance and health benefit, to the participants recent international norms seem to have moved towards an ethical obligation for researchers to disclose all research results meeting these requirements (Knoppers et al. 2006). One example is the obligation stipulated in the Council of Europe's Additional Protocol to the Convention on Human Rights and Biomedicine, concerning Biomedical Research. The Additional Protocol came into force in 2007 and is legally binding for countries that have ratified it. It is the reflection of the human rights based approach of the Council of Europe and stipulates a right to know:

> Article 27. If research gives rise to information of relevance to the current or future health or quality of life of research participants, this information must be offered to them. (COE 2005)

This stance is affirmed by the OECD guidelines, which take informing the study participant as the default version, of which the individual can opt out.

> 4.14 In certain circumstances, as permitted by applicable law and the appropriate authorities, where the participants may be provided with feedback of individual-level results arising from research, the operators of the HBGRD should provide clear information to the participant of the consequences of receiving such results and should inform the participant of their right to opt out from receiving such results. Non-validated results from scientific research using an HBGRD's human biological materials and data should not be reported back to the participants and this should be explained to them during the consent process.

Indeed, a provision, and especially an obligation, to return individual research results is seen by biobank managers and researchers as a significant burden that might hamper research or at least render projects more expensive while being of questionable benefit. Solutions that are in open conflict with the strong tradition of the human rights framework in Europe are not likely to be collectively acceptable nor is it desirable to undermine the strong focus of citizens' rights. Ethical positions that motivate a general practice of not returning results (Forsberg, Hansson, and Eriksson 2009) and lobby for a "shift of focus from autonomy and individual rights toward collective responsibility and solidarity" are not even necessarily in the interest of researchers and science. At present, article 27 of the Additional Protocol does not create a significant burden on researchers: It is compatible with a practice of not returning results in the majority of cases of biobank research, because this research does not generate results that are of "relevance to the current or future health or quality of life of research participants". It is unlikely that RECs will approve a non-return-results policy if a project generates this type of results. In addition, trust of research participants and society could be significantly undermined – with a reduced willingness to participate in biobanks as a likely consequence – if biobanks adopted a rigid approach of not feeding back results under any conditions.

The three major themes outlined above are not the only issues that are discussed in the international literature. With their focus on privacy concerns of citizens in (mostly rich) countries they are certainly most pertinent to European policy-making today. However, in the future, other concerns remain to be addressed, among them issues such as collective consent or benefit sharing, particularly in the context of research carried out with participation of resource-poor populations. If biobanks aim at international collaboration, negotiating fair conditions that aim to prevent the dangers of discrimination and exploitation will be a highly relevant task.

3.3.3 Legal Risks

Although ethical issues are not fully resolved and controversy persists between different ethical traditions, recent guidelines have the potential to further harmonization. The guidelines provide enough options for biobank managers and researchers to choose a research friendly approach, while still granting sufficient protection of individuals who participate in biobank research. The crucial task is in the hands of international research and biobank organizations and networks. In order to use biobank resources in a responsible and efficient way – which implies facilitating international collaborations – they need to adopt the same options or at least mutually compatible technical and ethical frameworks. Although scholars have called repeatedly for a clarification of guidelines and the legal context (Deplanque et al. 2009), lessons learned from the past show that legal frameworks evolve slowly. Indeed, since in most countries the legal framework concerning biobanks remains poorly defined, uncertainty persists as to whether biobanks are taking legal risks, for example if they use broad consent. However, this is not likely to change within a short timeframe. In addition, any legal framework is always open to interpretation. Not infrequently, legislators have deliberately opted in favour of a somewhat vague legal framework, especially in areas that are rapidly changing, such as biotechnology. It may be assumed that if a biobank case is ever brought before a court the interpretation of present laws will take into account the directions provided by national or international guidelines. In Switzerland, for example, the federal court has taken into account guidelines of the Swiss Academy of Medical Sciences in order to clarify legally unresolved issues. In light of the absent, uncertain or patchy framework of different legal provisions internationally or even within the same country, the most efficient and pragmatic way forward for researchers and biobank managers is to choose a well-argued, harmonized framework in line with international guidelines. This approach could mean some legal risks, but these may be minimized if any regulatory uncertainties are explicitly addressed in formal contractual agreements (Goebel et al. 2010). In order to favour harmonization, template contractual agreements should be proposed by international networks involved in biobank research.

3.4 Conclusions

Proponents of biobank research promise personalized diagnostic and therapeutic approaches and public health benefits through a better understanding of the interactions between genes and the environment. Although such promises need to be taken with a grain of salt, the remarkable potential of biobanks as a research tool is uncontroversial. A realistic view is important. Researchers and private investors are at risk to exaggerate the potential in order to obtain funding. The tension between the true potential of biobanks and the ubiquitous hype is not beneficial for the scientific endeavor. If too much is promised public trust is undermined and valid future projects could be hampered.

Collaborations, at a national as well as an international level, are an indispensable strategy to maximize the benefit of biobanks. The necessary harmonization of technical procedures and ethico-legal provisions is in the interest of all stakeholders: it will foster the efficiency of research as well as the global protection of research participants. Without adequate protection and fair, transparent standards it is unlikely that the public – individuals, communities, populations – will be able to provide the trust and endorsement biobank research needs for its advancement.

The potential of biobank research is highly dependent on efficient solutions for the regulatory challenges. Europe can only take advantage of the wealth of information contained in its collections of samples and data if the ethical debate about research involving biobanks is adequately resolved. The main goal of this debate is to ensure the protection of the rights of those who have provided the human material, without unduly hampering research. International guidelines provide at present sufficient options to achieve this goal.

Concerning the three major issues discussed in this chapter – informed consent, privacy, and returning results to participants – a consensus is evolving towards (1) more acceptance of broad consent, if it comes with additional safeguards in the form of suitable oversight mechanisms, and (2) requiring researchers to offer informing research participants about individual study results under certain conditions.

The challenge for researchers, biobank managers, biobanks participants and society is today to choose and adhere to a harmonized approach in line with international guidelines, even if – or rather, because – the legal frameworks in different countries remain vague and open to interpretation.

References

Abbott, A. 2003. "With Your Genes? Take One of These, Three Times a Day." *Nature* 425 (6960): 760–62.

Andorno, R. 2006. "Biobancos poblacionales: un analisis juridico a partir de las experiencias islandesa y Estonia" [Population Bio-Banks: A Juridical Analysis Based on Icelandic and Estonian Experience]. *Law and the Human Genome Review* 25: 55–76.

Asslaber, M., and K. Zatloukal. 2007. "Biobanks: Transnational, European and Global Networks." *Briefings in Functional Genomics and Proteomics* 6 (3): 193–201.

Ballantyne, C. 2008. "Report Urges Europe to Combine Wealth of Biobank Data." *Nature Medicine* 14 (7): 701.

Bardelli, S. 2010. "Stem Cell Biobanks." *Journal of Cardiovascular Translational Research* 3 (2): 128–34.

Belmont Report. 1979. "The National Commission for the Protection of Human Subjects of Biomedical and Behavioral Research." The Belmont Report Ethical Principles and Guidelines for the Protection of Human Subjects of Research. April 18, 1979. Accessed March 4, 2011. http://ohsr.od.nih.gov/guidelines/belmont.html

Bevilacqua, G., F. Bosman, T. Dassesse, H. Hofler, A. Janin, R. Langer et al. 2010. "The Role of the Pathologist in Tissue Banking: European Consensus Expert Group Report." *Virchows Archiv* 456 (4): 449–54.

Botling, J., K. Edlund, U. Segersten, S. Tahmasebpoor, M. Engstrom, M. Sundstrom et al. 2009. "Impact of Thawing on RNA Integrity and Gene Expression Analysis in Fresh Frozen Tissue." *Diagnostic Molecular Pathology* 18 (1): 44–52.

Bovenberg, J., T. Meulenkamp, E. Smets, and S. Gevers. 2009. "Biobank Research: Reporting Results to Individual Participants." *European Journal of Health Law* 16 (3): 229–47.

Butler, D. 2010. "Human Genome at Ten: Science after the Sequence." *Nature* 465 (7301): 1000–1.

Cardoso, S., L. Valverde, A. Odriozola, X. Elcoroaristizabal, and M. M. de Pancorbo. 2010. "Quality Standards in Biobanking: Authentication by Genetic Profiling of Blood Spots from Donor's Original Sample." *European Journal of Human Genetics* 18 (7): 848–51.

Caulfield, T., and C. Weijer. 2009. "Minimal Risk and Large-Scale Biobank and Cohort Research." *Health Law Review* 17 (2–3): 53–58.

Chabannon, C., A. Honstettre, L. M. Daufresne, P. M. Martin, C. Bonnetaud, I. Birtwisle-Peyrottes et al. 2010. "Publication of Biological Samples Collections Catalogues by Tumor Banks." *Bull Cancer* 97 (2): 181–89.

Chung, T. H., J. H. Yoo, J. C. Ryu, and Y. S. Kim. 2009. "Recent Progress in Toxicogenomics Research in South Korea." *BMC Proceedings* 3 (Suppl. 2): S6.

Clement, B., G. Chene, and F. Degos. 2009. "A National Collection of Liver Tumours: Lessons Learnt from 6 Years of Biobanking in France." *Cancer Letters*.

COE. 2005. "Additional Protocol to the Convention on Human Rights and Biomedicine, Concerning Biomedical Research Strasbourg," 25.I.2005. Accessed March 2, 2011. http://conventions.coe.int/Treaty/en/Treaties/Html/195.htm

COE. 2006. "Council of Europe. Rec(2006)4 of the Committee of Ministers to Member States on Research on Biological Materials of Human Origin." Accessed March 2, 2011. https://wcd.coe.int/ViewDoc.jsp?id=977859

Data Protection Working Party. 2007. "Opinion 4/2007 on the Concept of Personal Data." Article 29, WP 136. Brussels: European Commission. Accessed March 2, 2011. http://ec.europa.eu/justice_home/fsj/privacy/docs/wpdocs/2007/wp136_en.pdf

Day, J. G., and G. N. Stacey. 2008. "Biobanking." *Molecular Biotechnology* 40 (2): 202–13.

Deplanque, D., G. Birraux, P. H. Bertoye, and E. Postaire. 2009. "Collections of Human Biological Samples for Scientific Purposes. Why Do Current Regulation Need to be Clarified and How?" *Therapie* 64 (4): 259–67.

Diez, D., A. M. Wheelock, S. Goto, J. Z. Haeggstrom, G. Paulsson-Berne, G. K. Hansson et al. 2010. "The Use of Network Analyses for Elucidating Mechanisms in Cardiovascular Disease." *Molecular Biosystems* 6 (2): 289–304.

Durham, N. P., and A. S. Hall. 1999. "Genes and the Heart: A Quest for New Therapeutic Strategies. Cambridge Healthtech Institute's First Cardiovascular Genomics Conference. Orlando, FL, USA, 11–12 January 1999." *Molecular Medicine Today* 5 (5): 195–97.

Ebner, A., J. R. Thyrian, A. Lange, M. L. Lingnau, M. Scheler-Hofmann, D. Rosskopf et al. 2010. "Survey of Neonates in Pomerania (SNiP): A Population-Based Birth Study – Objectives, Design and Population Coverage." *Paediatric and Perinatal Epidemiology* 24 (2): 190–99.

Elger, B., N. Biller-Andorno, A. Mauron, and A. M. Capron (Eds.). 2008. *Ethical Issues in Governing Biobanks: Global Perspectives*. London: Ashgate.

Elger, B. S. 2010. *Ethical Issues of Human Genetic Databases: A Challenge to Classical Health Research Ethics?* London: Ashgate.

Elger, B. S., and A. L. Caplan. 2006. "Consent and Anonymization in Research Involving Biobanks: Differing Terms and Norms Present Serious Barriers to an International Framework." *European Molecular Biology Organization Reports* 7 (7): 661–66.

Ennis, D. P., G. P. Pidgeon, N. Millar, N. Ravi, and J. V. Reynolds. 2009. "Building a Bioresource for Esophageal Research: Lessons from the Early Experience of an Academic Medical Center." *Diseases of the Esophagus*.

Enserink, M. 1998a. "Opponents Criticize Iceland's Database." *Science* 282 (5390): 859.

Enserink, M. 1998b. "Physicians Wary of Scheme to Pool Icelanders' Genetic Data." *Science* 281 (5379): 890–91.

European Science Foundation. 2008. "Population Surveys and Biobanking." Accessed March 2, 2011. http://www.esf.org/publications/science-policy-briefings.html

Forsberg, J. S., M. G. Hansson, and S. Eriksson. 2009. "Changing Perspectives in Biobank Research: From Individual Rights to Concerns about Public Health Regarding the Return of Results." *European Journal of Human Genetics*.

Garcia-Merino, I., N. de Las Cuevas, J. L. Jimenez, A. Garcia, J. Gallego, C. Gomez. et al. 2010. "Pediatric HIV BioBank: A New Role of the Spanish HIV BioBank in Pediatric HIV Research." *AIDS Research and Human Retroviruses* 26 (2): 241–44.

Goebel, J. W., T. Pickardt, M. Bedau, M. Fuchs, C. Lenk, I. Paster et al. 2010. "Legal and Ethical Consequences of International Biobanking from a National Perspective: The German BMB-EUCoop Project." *European Journal of Human Genetics* 18 (5): 522–25.

Gurwitz, D., I. Fortier, J. E. Lunshof, and B. M. Knoppers. 2009. "Research Ethics: Children and Population Biobanks." *Science* 325 (5942): 818–19.

Gurwitz, D., J. E. Lunshof, G. Dedoussis, C. S. Flordellis, U. Fuhr, J. Kirchheiner et al. 2005. "Pharmacogenomics Education: International Society of Pharmacogenomics Recommendations for Medical, Pharmaceutical, and Health Schools Deans of Education." *Pharmacogenomics Journal* 5 (4): 221–25.

Gurwitz, D., and M. Pirmohamed. 2010. "Pharmacogenomics: The Importance of Accurate Phenotypes." *Pharmacogenomics* 11 (4): 469–70.

ISBER. 2009. International Society for Biological and Environmental Repositories. Accessed March 2, 2011. http://www.isber.org/

Jiang, C. Q., T. H. Lam, J. M. Lin, B. Liu, X. J. Yue, K. K. Cheng et al. 2009. "An Overview of the Guangzhou Biobank Cohort Study-Cardiovascular Disease Subcohort (GBCS-CVD): A Platform for Multidisciplinary Collaboration." *Journal of Human Hypertension* 24 (2): 139–50.

Johnsen, I. K., S. Hahner, J. J. Briere, A. Ozimek, A. P. Gimenez-Roqueplo, C. Hantel et al. 2009. "Evaluation of a Standardized Protocol for Processing Adrenal Tumor Samples: Preparation for a European Adrenal Tumor Bank." *Hormone and Metabolic Research* 42 (2): 93–101.

Jones, D. A., and C. MacKellar. 2009. "Consent for Biobank Tissue in Somatic-Cell Nuclear Transfer." *Lancet* 374 (9693): 861–63.

Kaiser, J. 2002. "Biobanks: Population Databases Boom, from Iceland to the U.S." *Science* 298 (5596): 1158–61.

Kapeu, A. S., T. Luostarinen, E. Jellum, J. Dillner, M. Hakama, P. Koskela et al. 2009. "Is smoking an independent risk factor for invasive cervical cancer? A nested case-control study within Nordic biobanks." *American Journal of Epidemiology* 169 (4): 480–488.

Kavikondala, S., C. Q. Jiang, W. S. Zhang, K. K. Cheng, T. H. Lam, G. M. Leung et al. 2010. "Intergenerational 'Mismatch' and Adiposity in a developing population: The Guangzhou Biobank Cohort Study." *Social Science and Medicine* 70 (6): 834–43.

Knoppers, B. M., Y. Joly, J. Simard, and F. Durocher. 2006. "The Emergence of an Ethical Duty to Disclose Genetic Research Results: International Perspectives." *European Journal of Human Genetics* 14 (11): 1170–78.

Langseth, H., T. Luostarinen, F. Bray, and J. Dillner. 2010. "Ensuring Quality in Studies Linking Cancer Registries and Biobanks." *Acta Oncology* 49 (3): 368–77.

Latourelle, J. C., N. Pankratz, A. Dumitriu, J. B. Wilk, S. Goldwurm, G. Pezzoli et al. 2009. "Genomewide Association Study for Onset Age in Parkinson Disease." *BMC Medical Genetics* 10: 98.

Lefrere, J. J., and N. Coudurier. 2009. "Biobanks and Blood Transfusion in France: A Tool for Public Health." *Transfusion Clinique et Biologique* 16 (2): 148–51.

Lemonick, M. D. 2006. "The Iceland Experiment: How a Tiny Island Nation Captured the Lead in the Genetic Revolution." *Time* 167 (8): 50–51.

Matimba, A., J. Del-Favero, C. Van Broeckhoven, and C. Masimirembwa. 2009. "Novel Variants of Major Drug-Metabolising Enzyme Genes in Diverse African Populations and their Predicted Functional Effects." *Human Genomics* 3 (2): 169–90.

Meslin, E. M. 2010. "The Value of Using Top-Down and Bottom-Up Approaches for Building Trust and Transparency in Biobanking." *Public Health Genomics* 13 (4): 207–14.

Mitchell, R. 2010. "National Biobanks: Clinical Labor, Risk Production, and the Creation of Biovalue." *Science Technology and Human Values* 35 (3): 330–55.

Modin, C., B. Bjerregaard, T. Orntoft, and E. Hogdall. 2010. "Establishing the Danish CancerBiobank." *Ugeskrift for laeger* 172 (19): 1446–50.

National Bioethics Advisory Commission [NBAC]. 1999. *Research Involving Human Biological Materials: Ethical Issues and Policy Guidance, Report and Recommendations, vol. 1.* Rockville: National Bioethics Advisory Commission.

NCI. 2007. "Best Practices for Biospecimen Resources." Accessed March 2, 2011. http://www.allirelandnci.org/pdf/NCI_Best_Practices_060507.pdf

Netzer, C., and N. Biller-Andorno. 2004. "Pharmacogenetic Testing, Informed Consent and the Problem of Secondary Information." *Bioethics* 18 (4): 344–60.

Nutley, N. J. 2002. Roche and deCODE Unveil New Alliance in Drug Discovery and Development. Current News Releases January 29, 2002. Accessed March 2, 2011. http://www.datamonitor.com/store/news/roche_and_decode_unveil_new_alliance_in_drug_discovery_and_development?productid=2DEA0EE0-08B6-4FE7-97A3-EE821B8848DD

O'Brien, S. J. 2009. "Stewardship of Human Biospecimens, DNA, Genotype, and Clinical Data in the GWAS Era." *Annual Review of Genomics and Human Genetics.*

OECD. 2009. "Guidelines on Human Biobanks and Genetic Research Databases." Paris: OECD. Accessed March 2, 2011. http://www.oecd.org/document/12/0,3343,en_2649_34537_40302092_1_1_1_1,00&&en-USS_01DBC.html

OHRP. 2004. *Guidance on Research Involving Coded Private Information or Biological Specimens*. Rockville: Office for Human Research Protections.

Ottenhoff, T. H. 2009. "Overcoming the Global Crisis: "Yes, We Can", But Also for TB?" *European Journal of Immunology* 39 (8): 2014–20.

Posch, M. G., G. Gelbrich, B. Pieske, E. Lehmkuhl, C. E. Angermann, S. Stork et al. 2009. "The Biomaterialbank of the German Competence Network of Heart Failure (CNHF) Is a Valuable Resource for Biomedical and Genetic Research." *International Journal of Cardiology* 136 (1): 108–11.

Reed, W., and R. Bjugn. 2010. "Good Biobanks for Better Health." *Tidsskr Nor Laegeforen* 130 (11): 1156–58.

Riegman, P. H., B. W. de Jong, and A. Llombart-Bosch. 2010. "The Organization of European Cancer Institute Pathobiology Working Group and its Support of European Biobanking Infrastructures for Translational Cancer Research." *Cancer Epidemiology, Biomarkers and Prevention* 19 (4): 923–26.

Ritchie, M. D., J. C. Denny, D. C. Crawford, A. H. Ramirez, J. B. Weiner, J. M. Pulley et al. 2010. "Robust Replication of Genotype-Phenotype Associations Across Multiple Diseases in an Electronic Medical Record." *American Journal of Human Genetics* 86 (4): 560–72.

Rose, H. 2006. "From Hype to Mothballs in Four Years: Troubles in the Development of Large-Scale DNA Biobanks in Europe." *Community Genetics* 9 (3): 184–89.

Rudan, I., A. Marusic, S. Jankovic, K. Rotim, M. Boban, G. Lauc et al. 2009. " '10001 Dalmatians': Croatia Launches Its National Biobank." *Croatian Medical Journal* 50 (1): 4–6.

Rudloff, U., U. Bhanot, W. Gerald, D. S. Klimstra, W. R. Jarnagin, M. F. Brennan et al. 2010. "Biobanking of Human Pancreas Cancer Tissue: Impact of Ex-Vivo Procurement Times on RNA Quality." *Annals of Surgical Oncology* 17 (8): 2229–36.

Salvaterra, E., L. Lecchi, S. Giovanelli, B. Butti, M. T. Bardella, P. A. Bertazzi et al. 2008. "Banking Together: A Unified Model of INFORMED Consent for Biobanking." *European Molecular Biology Organization Reports* 9 (4): 307–13.

Schwartz, J. 1999. "For Sale in Iceland: A Nation's Genetic Code." *Washington Post*. January 12, 1999. Accessed March 4, 2011. http://www.highbeam.com/doc/1P2-572603.html

Secko, D. M., N. Preto, S. Niemeyer, and M. M. Burgess. 2009. "Informed Consent in Biobank Research: A Deliberative Approach to the Debate." *Social Science and Medicine* 68 (4): 781–89.

Talens, R. P., D. I. Boomsma, E. W. Tobi, D. Kremer, J. W. Jukema, G. Willemsen et al. 2010. "Variation, Patterns, and Temporal Stability of DNA Methylation: Considerations for Epigenetic Epidemiology." *The FASEB Journal*.

Teunissen, C. E., A. Petzold, J. L. Bennett, F. S. Berven, L. Brundin, M. Comabella et al. 2009. "A Consensus Protocol for the Standardization of Cerebrospinal Fluid Collection and Biobanking." *Neurology* 73 (22): 1914–22.

Vlastos, F., S. Lacomme, P. Wild, S. Poulain, J. Siat, G. Grosdidier et al. 2009. "Do evolving practices improve survival in operated lung cancer patients? A biobank may answer." Journal of Thorac Oncology 4 (4): 505–511.

Williams, R., C. Mant, and J. Cason. 2009. "The Infectious Diseases BioBank at King's College London: Archiving Samples from Patients Infected with HIV to Facilitate Translational Research." *Retrovirology* 6: 98.

WMA. 2008. "WMA Declaration of Helsinki – Ethical Principles for Medical Research Involving Human Subjects." 59th WMA General Assembly, Seoul, October 2008. Accessed March 4, 2011. http://www.wma.net/fr/20activities/10ethics/10helsinki/

Womack, C., and N. M. Gray. 2009. "Banking Human Tissue for Research: Vision to Reality." *Cell and Tissue Banking* 10 (3): 267–70.

Yuille, M., K. Dixon, A. Platt, S. Pullum, D. Lewis, A. Hall et al. 2009. "The UK DNA Banking Network: A 'Fair Access' Biobank." *Cell and Tissue Banking* 11 (3): 241–51.

Zika, E., D. Paci, A. Braun, S. Rijkers-Defrasne, M. Deschenes, I. Fortier et al. 2010. "A European Survey on Biobanks: Trends and Issues." *Public Health Genomics*.

Chapter 4
A Sense of Entitlement: Individual vs. Public Interest in Human Tissue

Nils Hoppe

4.1 Introduction

The arena of the provision of health services and products is one where the contrast between what can be supplied and what is required is particularly stark. The more scarce the health good in question, the more pressing the issue of procurement and provision to the patient becomes. This was particularly visible when the recent H1N1 flu scare caught large parts of Europe unawares and the pharmaceutical industry was struggling to meet the estimated demand for vaccines, which caused a public debate on manufacturing delays (see, e.g., Ruiz 2009). The scarcity in the case of H1N1 vaccines was the result of an acute and unexpected hike in demand. It seems clear that a number of causes other than an unexpected pandemic may result in problems in the ability to provide some therapeutic interventions to patients. The more innovative a therapy, the more likely it is that there will be difficulties in meeting a broad demand. This was and is true for organ transplantation and is also the case in the context of human tissue-derived products. In Germany, some experts estimate that the current annual therapeutic demand for tissue donations amounts to approximately 8,000 corneae, 800 heart valves and 500 vascular grafts; at the same time, the actual transplantation figures for 2009 leveled off at 6,000 corneae and 500 heart valves and vascular grafts (combined) (DGFG 2009, 7). In the UK, figures from 2007 showed an acquisition of 4,021 corneae for transplantation, only 2,512 of which were suitable for transplantation (Transplant UK 2007, 32). The shortage of corneae extends from the therapeutic context to the research context, where a commercial stimulus to counter shortage is being discussed openly. Curcio (2006, 2748) raises the issue of pricing in the acquisition of human-derived material for research:

> It should by now be clear that although freely given, donor eye tissues for research are not harvested, processed and delivered for free.

N. Hoppe (✉)
Institute for Philosophy, CELLS – Centre for Ethics and Law in the Life Sciences,
Leibniz Universitaet, Hannover, Germany
e-mail: nils.hoppe@agmb.uni-hannover.de

C. Lenk et al. (eds.), *Biobanks and Tissue Research*, The International Library of Ethics, Law and Technology 8, DOI 10.1007/978-94-007-1673-5_4,

Whilst the vicissitudes of the market seem to play a role in the downstream handling of human-derived material, they are generally excluded from the discussion at donation stage. I will address this issue briefly below. Further, a complicating factor in terms of human tissue-derived products is clearly that the availability of the raw material for these products is on the one hand dependent on an extremely onerous information-and-consent-based system and that, on the other hand, those manufacturing these products cannot influence the availability of the raw material.

This chapter works on two assumptions which I will attempt to make plausible in due course. Owing to the brevity of this chapter, I cannot do justice to the complexity of an appropriate justification but will point to appropriate literature which has provided these. The assumptions are that:

1. developing and providing human-derived material[1] for therapeutic purposes is desirable from a societal perspective; and
2. it is in society's interest to protect and respect the material's source and his or her wishes in relation to the procurement of the material (though not necessarily all the time).

I will then try to demonstrate that it is legally and ethically justifiable to make a clear distinction between the normative requirements for procurement from living sources and from cadaveric sources, which has a particular impact on issues of consent. In particular, I am going to argue that where there is an overwhelming demand for cadaveric material, and the two assumptions outlined above are true, it is legally and ethically justifiable to tip the scales in favour of the public interest and against the individual in certain cases. This will result in the proposition of a three-tiered model: where the availability of certain human-derived material is life-saving the public interest outweighs individual rights; where the availability of the material would improve the health of another, the public interest is great but donation and procurement should be altruistic and voluntary; and where the objective of the procurement is commercial research, it should be left to the donor to decide whether to contribute to the research on an altruistic basis or for a reward.

4.2 The Societal Perspective

My first assumption is that the development and provision of human-derived products for therapeutic purposes is a desirable activity. This will be the case where the benefits of the products in question substantially outweigh the detriments of procuring or using the products. I will turn to the benefits first.

[1] I deliberately make no clear distinction between organs for transplantation and other tissues. Instead, my distinction will centre on whether the material has life-saving potential or not.

There is substantial evidence that these kinds of products possess characteristics which cannot in all cases be compensated for by means of recourse to animal or synthetic models. A very current and topical example is the development of innovative heart valve replacements, based on a decellularised collagen matrices derived from donor[2] heart valves. The chemical removal of all vital donor cells leads to a mechanically functional heart valve replacement which integrates seamlessly and seemingly without adverse immunological reaction into the recipient organism (Cebotari et al. 2006). Neither conventional donor heart valves nor xenografts or artificial heart valves reach the level of superiority currently being demonstrated by these decellularised valves. The example can be extended to the use of human tissue grafts for the treatment of burns victims and to the transplantation of corneae for the benefit of patients with poor eyesight. The complete spectrum from health-improving to life-saving treatments is covered by human tissue-derived products which cannot easily be replaced by equally good alternatives.

The question why we would need to replace them at all if they are available from human donors is the result of a general, and in some cases rather diffuse, discussion of the detriments of making the human body available for further use after death. There is an assumption that where an alternative is available we should opt for that alternative rather than make the most of those "human resources" theoretically available to us. This reluctance or uneasiness about making the body available has many cultural, moral and legal bases which have been described in great detail elsewhere (Böhnke 2010; Hoppe 2009; Steinmann et al. 2009; Dickenson 2008; 2007; Hardcastle 2007; Nwabueze 2007; Wilkinson 2003) and which, in many cases, amount to little more than irrational and occult reflexes. Two general objections can be characterised as a recurring theme in this discussion.

One is the assertion that the human body should not be merely a means to an end (for a more detailed discussion see Hoppe 2009, 5–9). This position is summarised, though not adequately justified, in the Warnock Report:

> That people should treat others as a means to their own ends, however desirable the consequences, must always be liable to moral objection. Such treatment of one person by another becomes positively exploitative when financial interests are involved. (Warnock, M. et al. "Report of the Committee of Enquiry into Human Fertilisation and Embryology", section 8.17, quoted in Wilkinson 2003, 35)

The problem with this objection to making the body available is that this discussion format is often deployed in cases where the teleological development of an argument is at stake. Janet Radcliffe-Richards writes persuasively:

> This format usually encourages protagonists to collect into an unsorted heap whatever arguments look as though they might have any persuasive force on their side, and because

[2] I will adhere to the donor/donation terminology even though I disagree with the legal implications this has. Technically, a donation is a property transfer. As long as the source of the material is said to have no property interest but is entitled to transfer that non-existent property interest to another, the terminology used is at best incomplete and at worst deliberately inappropriate. James Harris makes this point when he distinguishes between full-blooded ownership and mere property (Harris 1996, 28–29).

> people may be on the same political side for different moral reasons, or have the same moral principles but reach different political conclusions, the political arguments tend to obscure both the real issues and the logical structure of the controversy. (Radcliffe-Richards 2003, 139–40)

Such a token application of the "no-means-to-an-end" argument generally whittles down its usefulness in debate and tends to fail to adequately answer the question why we cannot use cadavers as means to an end (the often neglected distinction between live and post-mortem procurement will be made below). The main criticism of this objection must be that its use seems to no longer require elaboration, application and justification. Merely the assertion that individuals should not be used as means to an end seems to be sufficient for those availing of this objection. In the context of making decisions relating to plainly life-saving necessity, this type of argument – delivered in this way – is of very little assistance.

The Warnock quote, above, builds the bridge to the other objection, which centres on the prevention of exploitation (for an extensive discussion of this, see e.g. Wilkinson 2003, 9–55). I have discussed the second objection to using the human body for fear of exploitation before (Hoppe 2009, 137–38; 2010) and will therefore only give a very brief version of my criticism here. The obligatory application of notions of donation whilst at the same time denying the source any semblance of a property right does not serve to increase the protection of the source's rights. Rather, it decreases the protectable entitlements the source may have in their own body, whilst at the same time accepting that third parties who have come into possession of the source's material are free to invest some time and effort to create property capable of being transferred. The only protagonist who has been excluded from what is doubtless an exploitation chain is the source whom we claim to be protecting with the current system. The system therefore does not prevent exploitation, it merely prevents self-exploitation: I may not benefit from the fruits of my own body but others may. Or, to misapply the first objection above: others may use my body as a means to their ends, but I may not use my own body as a means to my own ends.

These two standard objections are simply insufficient to justifiably reject using the human body or its products for the benefit of others per se. This means that, in principle, the kinds of tissue we may need to produce an innovative therapeutic product should be available and should be used. Their development, production and provision relieve suffering in seriously ill patients and may well reduce the financial burden on society as a whole which would otherwise have to provide ongoing and expensive healthcare for these patients. Having, admittedly somewhat cursorily, done away with the two main objections to the fundamental availability of the human body we must ask the question how and when the body is available. Simply because we make the human body available does not mean that this is so without limitations. It is my argument that the strength of these limitations is dependent entirely on the exact context of supply and demand.

4.3 Respect for the Individual

The second assumption I made is that it is in society's interest to protect and respect the material's source and his or her wishes in relation to the procurement of the material. I do not want to focus this part on the trivial discussion of a person's right to bodily integrity and thus his or her right to determine what is procured from his body. I rather want to demagnify the issue to the more abstract question of why it is necessary to maintain a system of respect for individuals in the context of tissue procurement. The question may sound frivolous but we are discussing this issue in an environment where the courts have repeatedly asserted in all manner of legally sound and unsound ways that sources have no property in their own body when they are alive.[3] At the same time, it is also regularly asserted that there should be some sort of residual respect for the wishes of the individual when determining what is to be done with material after his or her death. The two ideas seem, from a legal point of view, somewhat conflicting. Why should I have a better control right when I am no longer capable of being a holder of rights? Either there is a control right or there is not. Either I can make binding inter vivos arrangements for the handling of my body and its parts or I cannot.

The interesting question seems to be in relation to whether our notion of respect for the individual survives the individual's death. In fact, the distinction between living and cadaveric donors often does not receive appropriate attention. I will briefly draw distinctions here and lead on to a discussion of why the individual interest may well outweigh in the live donation scenario and why the public interest may outweigh in the post mortem scenario. I have criticised the unclear differentiation of whether material is taken from a living donor or from a post-mortem donor before and have endeavoured to provide a differentiation matrix to assist in categorising the debate (Hoppe 2009, 15). This is because failing to be exact about some pivotal questions in this discussion renders any comparison void.

> One author might speak of issues in relation to live organ donation whilst another addresses the retention of children's organs after post-mortem examinations. [...] A slight but significant difference in their premise (such as between the taking of tissue from live and from post-mortem sources) is sufficient to totally unhinge the comparability of their logic and their analysis. (Hoppe 2009, 24)

As an illustration, the regularly cited Moore case is all too often discussed in the same breath as the Alder Hey Scandal. In Moore, two physicians dishonestly extracted cells from a patient over many years in order to create a cell-line culminating in a patent and profit. In Alder Hey, an overextended pathologist misinterpreted institutional policy and accumulated a gruesome collection of body parts taken during post-mortem examinations and autopsies. John Harris (2002, 546) puts his finger on it when he writes about sentiments in the context of Alder Hey:

---

[3] The classic cases of Moore (793 P 2d 479), Catalona (437 F Supp 2d 985) and Greenberg (264 F Supp 2d 1064) are usually cited here. For a discussion of all three, and the more exotic decision in Yearworth ([2009] EWCA Civ 37), see Hoppe (2009, 107–15).

> A quite absurd, if understandable, preoccupation with reverence and respect for bodily tissue has come to dominate discussions of retained tissues and organs in the wake of the Alder Hey revelations. We do not normally feel this reverence for our bodily remains, tissue and organs when alive – why suddenly this morbid post mortem preoccupation?

A living patient whose genetic makeup was abstracted in order to further another's financial interests cannot be discussed using the same arguments as the living parents of a deceased toddler whose heart was immorally but legally kept by the hospital which procured it in accordance with the law in the first place. It appears clear that a distinction must be made in this context. For the purposes of this chapter, the distinction should be that the type of respect extended to a living donor must differ to the type of respect extended to a cadaveric donor.

4.4 Commercialisation

At this point of the debate, we are still not making a sound distinction between the procurement from a living donor and that from a post-mortem source of tissues and cells (which will become intolerably difficult towards the end of this section). In this part, I will attempt to demonstrate that the wholesale arguments usually used to underpin the current system of altruistic donation pursuant to an individual's wishes during their lifetime do not apply in all cases and that a further distinction must be drawn between donation for life-saving treatment, donation for health-improving treatment and donation for research.

A superficially plausible line of reasoning is that it is imperative for the stability of a health care system heavily footed in solidarity to maintain a framework of altruism. For if the system gives individuals the feeling that their wishes carry no weight, they will not make their bodies available for therapy or research after their deaths[4] (which makes an exclusive reliance on altruistic donations rather burdensome in the context of desperately required materials). Sýkora sums up a position found in numerous instruments and commentaries (2009, 13):

> In the last four or five decades, a general consensus has been formed that paid and unpaid donation systems are mutually exclusive and society has to endorse ethically superior altruistic donations instead of bodily commerce. Furthermore, it is believed that practicing voluntary and unpaid medical donations has a prosocial effect on the society as a whole and therefore "a major argument for exclusive reliance on unpaid donation is that, unlike paid donation, it promotes altruism and social solidarity".

An example of this position finding its way into a legislative instrument is the preamble of the European Union's human tissue directive[5]:

---

[4] By, for example, making private arrangements for the body to be physically removed before any material can be taken.

[5] Directive 2004/23/EC of the European Parliament and of the Council of 31 March 2004 on setting standards of quality and safety for the donation, procurement, testing, processing, preservation, storage and distribution of human tissues and cells.

> As a matter of principle, tissue and cell application programmes should be founded on the philosophy of voluntary and unpaid donation, anonymity of both donor and recipient, altruism of the donor and solidarity between donor and recipient. Member States are urged to take steps to encourage a strong public and nonprofit sector involvement in the provision of tissue and cell application services and the related research and development. (2004/23/EC, Preamble, para. 18)

The whole preamble is in general very careful to avoid interfering with domestic notions of what is an individual and when life might begin, but has no qualms about dictating the doctrine of enforced altruism into domestic statute books (including, oddly, a reference to research and development, which is not within the ambit of the directive but may just be another illustration of poor differentiation). In this case, again, it is difficult to find an appropriate justification why altruistic giving is supposed to be ethically superior to commerce in all cases. The paradigm of altruistic giving is simply applied in too many cases without appropriate differentiation. Consider the following scenarios:

1. I insist that my estate is paid €10,000 for the right to take one of my organs or tissue to save a life;
2. I insist that my estate is paid €10,000 for the right to take my corneae to treat another's failing eyesight; or
3. I require a personal payment of €10,000 for the right to use my body substances for research leading to a commercially viable product.

In the case (1), there probably is a good argument that my priorities are ethically dubious (without delving into the depths of the hedonistic paradox and other discussions of putting material gain before saving another's life, I am comfortable with asserting that this would be clearly morally objectionable) and the activity thus unacceptable. In case (2) it may well be that a distinction in my estate's favour could be made on the basis that the proposed use is not life-saving. "Only" the recipient's general well-being will be at stake, rather than his life, which opens the equation up to an exercise of balancing entitlements. Even if we do not subscribe to the pecuniary aspect, it would certainly be thinkable that my desire to remain intact after death can outweigh the other's desire to see again. In case (3) many people would suggest that there is a reasonable argument in my favour. On the basis of the discussion of the issue of exploitation, above, it is simply not plausible that everyone else is entitled to profit from my body, yet I am not. Even in the case of Moore (where profit-sharing between donors and researchers was deemed to be contrary to the public interest) a minority argument was put forward that it may well be in the interests of the donor – and of society – if the donor were permitted to participate in the pecuniary proceeds of the research:

> [...] the person [who furnishes the tissue] should be justly compensated. [...] If biotechnologists fail to make provision for a just sharing of profits with the person whose gift made

it possible, the public's sense of justice will be offended and no one will be the winner. (Thomas Murray, quoted by Mosk, J. in Moore[6] at para. 3, near note 21)

In fact, in our scenario (3) and in the Moore case, there is simply nothing wrong with me saying that I would not make my cells available for free, but would be willing to undergo the onerous and potentially risky procedure of procurement for a payment as I can autonomously dictate which benefit threshold ought to be crossed before I consent to a detriment. So the public interest argument usually used to abnegate an individual's ability to participate in commercial exploitation can be turned on its head and a plausible justification for profit-sharing can be shaped. It is not necessarily against the public interest to remunerate a donor – on the contrary. Where the donor's cells are required for important research and he is only willing to sell them for a reasonable amount, it would be contrary to the public interest to decline the purchase on the grounds that he ought to give them altruistically. Further, there is a persuasive line of reasoning which reduces the question of commercialisation to one of reciprocity detached from money – or as contract lawyers would put it: to one of a benefit or a detriment to both parties. Sýkora (2009, 16) points out some commentators which field an objection to the provision of any benefit to encourage donation of body products (such as time off work, free travel to the place of donation, etc). Such benefits would negate the altruistic character of the giving and the general objection against commercialisation becomes not one of market-exchange but simply an objection against money (see also Hoppe 2009, 130–32). Whilst the question whether this is an issue of restitution or one of reward remains largely unaddressed, there is little wrong with this criticism of the objection against commercialisation where it is levelled merely at the issue of money rather than at an exchange.

In summary, it seems that there are certain circumstances where the respect for the individual's wishes is paramount, and some circumstances where this is not the case. Additionally, there seem to be some circumstances where the commercialisation of human-derived material is inappropriate, and some circumstances where it may be appropriate. Returning full circle to our scenario (1) above, it appears at first glance that demanding payment in order to save another's life is ethically dubious. On simple balance, it seems that saving the other individual's life must be superior to any considerations for the respect of the (ex-)individual who has the means to save that life. As long as we do not harm an equally important right of the donor, we seem to be justified in letting necessity dictate our actions and to simply take the organ or tissue necessary to save the life. This would not work with a live donor (for the trivial reasons given at the outset) but it would be perfectly justified in the context of a cadaveric donor who has stipulated that his body would only be available for cash or not available at all.

[6]Moore v. The Regents of the University of California et al. 51 Cal.3d 120 (Supreme Court of California), 9 July 1990.

4.5 Individual vs. Public Interest

I have endeavoured to show above that there is an overwhelming public interest in making human tissue-derived therapies available to society. Were organs and tissues actually made available in sufficient quantities on the back of a system footed in voluntary giving, there would be no need for the discussion of alternative models of governance in the same way that there would be less need for road traffic regulation if individuals autonomously negotiated traffic in an appropriate way. The model I propose is based on combinations of the distinctions I have drawn above. Life-saving treatment deserves different models of governance to health-improving treatment (both are in the category "therapeutic"). Research deserves different models of governance to cosmetic treatments (both being in the category "non-therapeutic"). The type of model of governance for each of these categories can possibly be modelled on the weighting of public interest vs. individual interest and I will attempt to show this weighting in this final section.

I have painfully avoided touching on the issue of consent and have spoken of control rights so far. Exercising a control right necessarily goes hand in hand with consenting or refusing certain types of interaction with others and so includes the notion of consent to a certain extent. In the following quote, Mullen and Widdows (2009, 174) use the idea of consent to explain why we might respect stipulations in relation to a conditional gift:

> One aspect of the issue of consent is the ethical commitment to the donor having the power to decide the use to which their donation is put. Conditional gift appears to privilege the importance of respect for the preferences of the individual over any communal good (including perhaps the health needs of others) that could be gained through the use of the tissue. **This raises questions of the relative weight that should be given to individual preferences and against the interest of others**. (Emphasis added).

It seems true that respect for the individual may transcend that individual's demise if we give effect to his or her wishes *ex post*. The question seems to be whether this should be so in every case. The determining factor during the lifetime of an individual seems to be that we will happily respect their wishes as long as these wishes do not interfere with a superior interest. As a matter of public interest, we do not interfere with an individual's desire to practise his archery skills as long as he does not do so in a crowded playground. In most jurisdictions, we do not expect an individual to assist in an emergency if he would have to put his health or life at risk. If he can come to the rescue of another individual in need without putting his health or life at risk, we see the matter differently – again, as a matter of public interest. Why would we apply a different yardstick where the act of rescuing the other party entails giving material from my body? During my lifetime, this act of rescuing another is in many cases supererogatory and should be up to individual, voluntary decision-making. Once my health and life have deteriorated to the point of death, there is no sound argument why I cannot be compelled to assist.[7] For the reasons discussed, there seems to be a plausible argument why the individual interest outweighs the public interest in cases where the donor is still alive (and this does, in some cases, include the individual interest in commercialising his own body). In cases where the donor is dead, the scales may well be tipped in favour of the public interest, especially in cases where the material we need to procure is of a life-saving quality. In the final section of this chapter, I will try to describe and delineate the categorisations I have introduced above.

4.6 Conclusions

The three-tier categorisation contained in the sections above on the one hand relates to the nature and destination of the material we wish to procure and to the individual's and the public's right to control that material. The important distinction between live donation and post-mortem donation does not come into play in this categorisation as in this instance, the fundamental rights of the individual rightly negate any attempts to compel access to their bodily materials. It is clear, therefore, that I am not advocating the forceful removal of tissues from a living donor in any case.

In the context of post-mortem extraction of material, where the nature of the material is life-saving (as may be the case, inter alia, with complete organs, vascular grafts or heart valves), there is a plausible argument that the public interest in saving a life dwarves the individual's right to determine what may or may not be done with his body. I propose that where an individual can be identified whose life can be saved using the cadaveric tissue or organs of a potential donor, this should be done, even in the teeth of the individual's or his relatives' wishes. There seems to me to be no persuasive moral or legal argument for not interfering with a dead person's dignity (if there is such a thing) in cases where this would entail the death of a living person.

Where the tissue we wish to procure has merely health-improving character (such as, inter alia, corneal grafts), the current system of altruistic and voluntary donation can remain in place. The public interest in making these materials available does not outweigh the individuals' interests to the extent that the balance of control rights tips either way.

Finally, where the material is required for research or other non-therapeutic use, it can be left up to the individual to determine whether his or her material is made available free and altruistically or whether a reward of some nature (in the sense of Sýkora's reciprocity (see above)) is required. The categorisation might be summarised as shown in Table 4.1:

[7] Certainly not a new idea. See Spital and Erin (2002); Spital (2003, 2005a, b, 2006). But also see: McGovern (2002).

Table 4.1 Three-tiered approach to procurement governance

	Public interest	Individual interest	No choice	Altruistic donation	Reward
Life-saving material	Yes	No	Yes	No	No
Health-improving material	Yes	No	No	Yes	No
Non-therapeutic use	Yes	Yes	No	Yes	Yes

This conclusion is simply a further illustration of the fragmented nature of this topic of debate. The questions of commercialisation and control rights, for example, cannot be applied uniformly in all contexts. As I have outlined above, in cases where a life might be saved, holding the post-mortem organ ransom on the basis of ex ante wishes of the deceased seems inappropriate. Where no lives are immediately at stake, and commercially viable research is proposed, there can be parallel systems of altruism and reward.

The categorisation is crude and the proposition of no-choice in some scenarios dangerously provocative and certainly not made sufficiently plausible in this brief chapter. It should, however, serve to underline the significant point I seek to make: individual interests cannot in all cases be paramount in decision-making and a more balanced system of individual and public interests, with the possibly provocative outcome of a limitation of individual choice, should be seriously considered.

Acknowledgments I am very grateful to Jane Kaye, Liam Curren, John William Devine, Naomi Hawkins, Nadja Kannelopoulou and Karen Melham at the Centre for Health, Law and Emerging Technologies at Oxford for their hospitality and stimulation discussions during the drafting of the first versions of this chapter. As ever, all inaccuracies and errors remain my own.

References

Böhnke, O. A. 2010. *Die Kommerzialisierung der Gewebespende* [The commercialization of tissue donation]. Frankfurt am Main: Peter Lang.

Cebotari, S., A. Lichtenberg et al. 2006. "Clinical Application of Tissue Engineered Human Heart Valves Using Autologous Progenitor Cells." *Circulation* 114: I132–37.

Curcio, C. 2006. "Declining Availability of Human Eye Tissues for Research." *Investigative Ophthalmology & Visual Science* 47 (7): 2747–49.

Deutsche Gesellschaft für Gewebetransplantation. 2009. "Jahresbericht 2009." Hannover: DGFG, Accessed March 3, 2011. http://www.gewebenetzwerk.de/startseite/veranstaltungen/jahresbericht-2009/download.html

Dickenson, D. 2007. *Property in the Body – Feminist Perspectives*. Cambridge: Cambridge University Press.

Dickenson, D. 2008. *Body Shopping – The Economy Fuelled by Flesh and Blood*. Oxford: One World.

Hardcastle, R. 2007. *Law and the Human Body – Property Rights, Ownership and Control*. Oxford: Hart Publishing.

Harris, J. 2002. "Law and Regulation of Retained Organs: The Ethical Issues." *Legal Studies* 22 (4): 527–49.

Harris, J. W. 1996. *Property and Justice*. Oxford: Oxford University Press.

Hoppe, N. 2009. *Bioequity – Property and the Human Body*. Farnham: Ashgate.
Hoppe, N. 2010. Cui bono? Eigentum am eigenen Körper im internationalen Vergleich. Berliner Debatte Initial 21(4): 19–27.
McGovern, T. 2002. "Flawed Proposal for Universal Conscription of Cadaveric Organs Neglects Moral, Long-term, and Societal Implications." *American Journal of Kidney Disease* 36: 609–10.
Mullen, C., and H. Widdows 2009. "An Investigation of the Conception, Management and Regulation of Tangible and Intangible Property in Human Tissue: The PropEur Project." In *Altruism Reconsidered – Exploring New Approaches to Property in Human Tissue*, edited by M. Steinmann, P. Sýkora, and U. Wiesing, 169–79. Farnham: Ashgate.
Nwabueze, R. 2007. *Biotechnology and the Challenge of Property – Property Rights in Dead Bodies, Body Parts, and Genetic Information*. Aldershot: Ashgate.
Radcliffe-Richards, J. 2003. "Commentary: An Ethical Market in Human Organs." *Journal of Medical Ethics* 29 (3): 139–40.
Ruiz, R. 2009. "Behind the H1N1 Vaccine Shortage." *Washington Post*. October 30, 2009. Accessed March 3, 2011. http://www.forbes.com/2009/10/27/swine-flu-vaccine-lifestyle-health-h1n1-shortage.html
Spital, A., and C. Erin. 2002. "Conscription of Cadaveric Organs for Transplantation: Let's at Least Talk About it." *American Journal of Kidney Disease* 39: 611–15.
Spital, A. 2003. "Conscription of Cadaveric Organs for Transplantation: Neglected Again." *Kennedy Institute of Ethics Journal* 13: 169–74.
Spital, A. 2005a. "Conscription of Cadaveric Organs: We Need to Start Talking about It." *American Journal of Transplantation* 5: 1170–71.
Spital, A. 2005b. "Conscription of Cadaveric Organs for Transplantation: A Stimulating Idea Whose Time Has Not Yet Come." *Cambridge Quarterly of Healthcare Ethics* 14: 107–12.
Spital, A. 2006. "Conscription of Cadaveric Organs for Transplantation: Time to Start Talking about It." *Kidney International* 70: 607.
Steinmann, M., P. Sýkora, and U. Wiesing (Eds.). 2009. *Altruism Reconsidered – Exploring New Approaches to Property in Human Tissue*. Farnham: Ashgate.
Sýkora, P. 2009. "Altruism in Medical Donations Reconsidered: The Reciprocity Approach." In *Altruism Reconsidered – Exploring New Approaches to Property in Human Tissue*, edited by M. Steinmann, P. Sýkora, and U. Wiesing, 13–49. Farnham: Ashgate.
Transplant UK. 2007. *Transplant Activity in the UK*. London: The Stationery Office.
Wilkinson, S. 2003. *Bodies for Sale – Ethics and Exploitation in the Human Body Trade*. London: Routledge.

Chapter 5
Social Aspects of Biobanking: Beyond the Public/Private Distinction and Inside the Relationship Between the Body and Identity

Federico Neresini

The current literature exploring relationships between donor populations and biobanks is basically concerned with issues of privacy, confidentiality, consent, and regulation from a bioethical point of view (Mitchell and Waldby 2009). The purpose of this debate is to protect donor populations from unwarranted use of the information in various ways "extractable" from tissues – in most cases genetic information – and to ensure appropriate forms of regulation.[1]

Within this interpretative frame, biobanks are generally seen as structures of scientific research. They are constituted according to the logic and practices of scientific research and can therefore be wholly reduced to scientific projects and aims. What makes biobanks different from traditional herbariums, scientific pounds and experimental green-houses is the kind of material which is gathered, classified, preserved, and exploited. Because such material is part of a human body, it has an owner, and the information – especially genetic – that can be obtained from it reveals, or indeed affects, his or her identity. The advent of genetics, in fact, has strengthened the correspondence between body and identity, to the point that DNA can today be considered what the soul used to be in the past (Nelkin and Lindee 1995). This mystical and magical nature of genetic information means that, in June 2000, when the sequencing of the human genome was announced, we were told that the *Book of Life* had been read, the *Code of Codes* decoded, and the *Holy Grail* of human beings attained.

Notwithstanding this close link with human life, which is both promising and threatening at the same time, biobanks are considered scientific deposits for the purposes of scientific research and they can be insofar regarded as almost exclusively a matter for scientists.

According to this common point of view, therefore we have science on the one hand, and society on the other. This distinction is so deeply rooted in our culture that it is usually take for granted: science is a sort of separated world, a world

[1] See for example Haddow et al. (2007).

F. Neresini (✉)
PaSTIS Research Unit, Department of Sociology, University of Padua, Italy
e-mail: federico.neresini@unipd.it

C. Lenk et al. (eds.), *Biobanks and Tissue Research*, The International Library of Ethics, Law and Technology 8, DOI 10.1007/978-94-007-1673-5_5,

which creates the necessary conditions for the achievement of objective knowledge just thanks to this isolation from society. Mulkay defines this belief the "standard view of science" (1979), Latour calls it the "ready-made" science (1987), many others authors refer to it as a "positivistic" or "realistic" idea of science. However, it is possible to adopt a different perspective and consider biobanks as representing a paradigmatic case of the mutual constitution of the scientific and the social. Science and Technologies Studies refer to this in many ways: for example, in terms of "science in the making" (Latour 1987) or as the "co-production" of science and society (Jasanoff 2004) or again as "assembling humans and non-humans" (Latour 2005).

Framing the legal and ethical aspects of biobanks within the co-production of life sciences and the cultural and societal order – instead of seeking a better match between biobanks and their social context – means radically changing the basic problem from "How can we manage biobanking practices?" to "How has it been possible to have a society with biobanks?"

From a sociological point of view, in fact, biobanks are already deeply embedded in our society. But seeking to understand what is meant by saying that biobanks are an integral part of our society may yield insights useful for a debate on biobanks which concerns their legal and ethical aspects as well.

5.1 Inside and Outside the Laboratory

For our purposes, we can start with a very general question: what is needed to carry out scientific research? We could draw up a very long list, but here it is sufficient to reason with macro-categories. Therefore, what do we need for scientific research? Obviously, we need researchers, as well as buildings, tools and organisations, large sums of money, extensive knowledge, and an array of individual skills both cognitive and practical.

Moreover, scientific research consumes a great deal of "material", although this aspect tends to go unnoticed and remain in the background. If we think about the type of "material" used in the practice of scientific research, we can identify two main types:

(a) material which is created ad hoc for the laboratory (for instance, purified substances, especially modified organisms, accelerated particles);
(b) material directly taken from, or used in, the natural/social environment, which is therefore transformed into an open-air laboratory.

On observing how this material is processed by scientific laboratories, we realize that the conversion through which matter is made available for scientific research is a process full of mediations, conflicts and agreements. It is, that is to say, a *real* social process (Knorr-Cetina 1981; Latour and Woolgar 1979).

Firstly, to be borne in mind is that laboratories use objects which are not fixed entities that must be taken *as they are*. In fact, laboratories rarely work with objects as they occur in nature; instead, they work with images of objects, or with visual, auditory, electrical, etc., traces of them; they work with their components, their extractions or their "purified" versions. As Bruno Latour and Steve Woolgar write:

> it is not simply that phenomena *depend on* certain material instrumentation; rather, the phenomena *are thoroughly constituted by* the material setting of the laboratory. (1979: 64)

The artificial reality which scientists describe in terms of an objective entity has been constructed and mediated by the use of tools and devices. Science is therefore not just a theoretical enterprise, but a *worldly* one whose development depends on the ability to construct ad hoc instruments in order to produce ad hoc objects. This is exactly what occurs in the case of a biobank, a highly complex set of practices, technologies and conventions through which parts of the human body are made available for the work of scientists.

Secondly, science is constantly engaged in an attempt to transform the natural/social environment according to its needs; an endeavor which we can call the *laboratisation* of the world. Again, biobanks constitute a perfect example of this process: society must be organized so that human body specimens fulfilling the requirements of scientific research can be collected, preserved, and made available to the manipulations of scientists.

We should not fall into the trap of considering this tendency as a modern deviation from the traditional modus operandi of science. Science intrinsically considers the world to be a laboratory. Consider, for example, Louis Pasteur's work on the anthrax vaccine. In 1881, Pasteur needed to enrol a large number of farmers to validate his discovery. For this purpose he organized a public demonstration with two flocks of sheep. He only inoculated the sheep of the first flock, which survived, while the sheep of the second flock died. Pasteur thus managed to change the working practices of farmers so that they adhered to laboratory procedures such as disinfection, cleanliness, conservation, inoculation, timing and recording. Whilst Pasteur's public experiment was intended to convince the highest possible number of people to support his ideas and to gain new allies (Latour 1984), in other experimental situations it is a necessity imposed by the internal logic of experiments and the characteristics of their objects. The open-air nuclear tests conducted during the post-war period are the most immediate referent; but the same case is represented by the pendulum hung by Foucault from the vault of the Pantheon in Paris in 1851 to demonstrate the earth's rotation.

There is consequently plenty of proof to support the idea that laboratory walls have always been rather evanescent. However, the overlapping of science, technology and society has never been as evident as it is today. Biotechnology is an exemplary case. On one hand, in fact, research into genetically modified plants has required laboratory practice to be brought into the open, because it is not possible to verify the persistence of changes induced by genetic engineering until the new plants have been trialed in a "natural" environment.

On the other hand, attitudes and demands developed outside laboratories heavily influence their activity. This is the case of the debate on human embryo status, which has promoted research on adult stem cells and also explains why scientists have committed themselves to building genetically modified organisms which can be used for laboratory research, but are simultaneously able to avoid the limits posed by the bioethics debate. It was for this reason that Rudolf Jaenisch, one of the leading international experts on cloning, created in 2006 a biotechnological product able to produce embryonic stem cells but which cannot embed themselves in the uterus. As a consequence, researchers work with ethically compatible embryonic stem cells because these have been taken from an embryo which is not destined to grow into a foetus. This event is of great significance in itself; but the most interesting aspect is that it was the realisation of an idea publicly suggested some time ago by William Hurlbut, a bioethicist belonging to the Bioethics Committee set up by President Bush in 2004 with the explicit purpose of circumventing the ban on research with embryonic stem cells (Testa 2006: 155–59). In this case too, as in that of the biobank, determining where the laboratory's boundaries lie becomes very difficult, while their intrinsically social character becomes more and more visible.

Therefore, science and society has never been two distinct entities; they have always been mixed: on the one hand, science is part of society, an eminently social phenomenon; on the other, society is what it is partly because of science.

Biobanks perfectly embody this mixture of science and society: they collect, purify and conserve organic material (seemingly) from an environment external to that of science so that it becomes more available for scientific research. In doing so, however, they reorganise the environment according to their needs. All biobanks have this "hybrid" character, as is particularly evident in the case of national biobanks, which

> mediate between genetic information, biological samples, and patient experience on one hand, and between nation-states, populations, and "big science" on the other. (Mitchell and Waldby 2009: 2–3).

A biobank, in fact, can be usefully described as both a

> medical technology . . . [that is] . . . the various devices, instruments, and therapies used for diagnostic, therapeutic, rehabilitative, preventive, or experimental purposes as well as the practices and procedures associated with them. (Hogle 2008: 841)

and a "medical platform", that is,

> a specific combination of techniques, instruments, reagents, skills, constituent entities (morphologies, cell-surface markers, genes), spaces of representations, diagnostic, prognostic, and therapeutic indications, and related etiologic accounts. (Keating and Cambrosio 2003: 4)

But biobanks are, at the same time, more than medical platforms and more than medical technologies. They are also social institutions which are supported by, and which promote, social solidarity. They form a space for encounter between the promises of science and shared cultural goals such as health, the advancement of knowledge, economic development and national pride; they are the locus of tension between the collective good and private interests. As in the case of blood donation, national biobanks

> became woven into a myth of nationhood in which altruism and solidarity featured prominently. (Busby 2006)

Considering biobanks from the perspective in which science and society merge with each other therefore brings out various social and cultural implications which might otherwise remain largely invisible. It is obviously not possible to survey all these implications; here it will be sufficient emphasize that science and society are reciprocally transformed with particular reference to two general issues: on the one hand, redefinition of the distinction between public and private; on the other, transformation of the relationship between the body and identity.

5.2 Beyond the Public/Private Distinction

The famous case of the deCode project is very significant in this regard. The objectives of this project are ambitious: constructing a database with medical information regarding the Icelandic population, and connecting it to other population-wide databases, one with genetic data and the other about kinship relations, so that conditions associated with genetic anomalies can be identified.

According to the promoters of the deCode project, the relative geographical isolation of Iceland greatly reduced the possibility of the gene pool of the current population being mixed with that of other populations. Hence its variability was contained within limits suitable for such a profound and extensive genetic investigation.

But this "purity" was still not enough to transform this North European island into a giant laboratory; to do so, it was necessary to have reliable information relative to the genealogy of Icelanders, together with information on their state of health and that of members of the various generations of their families. Both conditions were progressively fulfilled through long and intense activity consisting of classifications, orderings, registrations and conservations; in short, a set of social practices which produced the necessary information and made it available. On the one hand, Iceland offers a wealth of genealogical information going back a thousand years; on the other, there is an archive of clinical files on every living Icelander and on their predecessors over a number of generations.

A law approved by the Icelandic parliament at the end of 1998 allowed the creation of a genetic database of the entire population based on the principle of silent-consent. When in 2001 the deCode researchers were granted a license by the government to use a similar gene pool, they were able to begin research by cross referencing the data from these three databases in order to determine the genetic origins of certain diseases. The deCode project has also been able to rely on another element of crucial importance: substantial financing made available partly by an agreement with the pharmaceuticals multinational Hoffman-Roche and partly by the listing of the deCode company on the stock exchange and the consequent issue of shares, many of which were purchased by the Icelanders themselves.

Today, almost 10 years after it began, the deCode project is in difficulties for various reasons: the presumed genetic "purity" of Icelanders is not as significant as was once believed; it is becoming increasingly clear that most diseases derive from a combination of genetic predispositions and environmental factors; and growing doubts have been expressed concerning the ethical admissibility of the criteria and means used to involve the Icelanders. The value of the shares has also collapsed and many small investors, especially Icelanders, have begun to support the few dissidents who had contested the deCode project from the outset.[2]

The case of deCode provides a good opportunity to observe how the practice of scientific research extends well beyond the usual boundaries of science, interacting with the environment as if it was a laboratory. An entire nation, its population, and its history have been the object of an experimental study. It was possible to begin the "laboratisation" of Iceland and of Icelanders given the existence of certain social-order characteristics; but once started, this process reshaped the same social order, remixing the economic structure, redefining boundaries which were the result of a centuries-long history, and destabilising acquired regulations. Consider, for example, the right to privacy. This long guaranteed a certain type of relationship between Icelandic citizens and public institutions, allowing the systematic collection and cataloguing of information on the family relationships and health of Icelanders; but this right has had to be completely revised in order to allow the use of such information for scientific research.

From a more general perspective, the *laboratisation* of Iceland and Icelanders has transformed one of the key principles of citizenship: the public interest. In fact, according to the logic of the genome bank, the "common good" simultaneously concerns

> population health benefits *and* commercial returns, national scientific prestige as well as global economic competitiveness. In other words, donor participation in biobanks contributes simultaneously to state *and* pharmaceutical interests, to public *and* private value. (Mitchell and Waldby 2009: 7)

This is where the overlap between the common good and the private interests supporting biobanks emerges: by promoting the objective of developing "drugs and diagnostics", it has presupposed presuppose

> the belief that the public good will be facilitated by commercial innovation of preventive, therapeutic, and diagnostic agents and tests. (Mitchell and Waldby 2009: 9)

Creation of the deCode project's biobank obviously raises numerous bioethical issues, most of which, as we have seen, are still awaiting satisfactory solution. But such issues are part of a process in which science and society co-evolve and transform each other. As they do so, they produce, among other effects, a redefinition of evaluation categories and criteria which hitherto seemed consolidated. Where lies the distinction between public and private interest in a case where donors are also shareholders, and are therefore the beneficiaries of any economic dividends yielded by the commercial enterprise to whose start-up their altruism has decisively contributed? What is meant by the expression "protection of privacy" if the boundaries between the owners and users of genetic information are constantly redrawn?

---

[2]For a discussion of the limits and difficulties of the deCode project see, among others, Palsson and Rabinow (1999, 2001), Sigurdsson (2001).

The term *biobank* seems to comprise the ideas of profit and therefore of private interest, on the one hand, and public utility and gratuitousness, on the other. But this twofold meaning is only apparently contradictory if we consider that the human body has become the raw material for the production of *biovalue* thanks to the development of scientific research. Both scientific and cultural developments have made the human body a valuable resource for the production of goods whose value resides in future payoffs in the form of new medical therapies or new knowledge with which to devise new therapies.

This process leads to the attribution of value to human tissues as materials useful for scientific research, and it adds new forms of the body's commodification and exploitation to the more traditional ones connected with work (Thompson 2005: 255–8). When the focus is on the production of *biovalue*, it becomes evident that once it has been acknowledged that scientific research cannot be confined within laboratory walls, the harvesting, conserving and transforming of human tissues – often, like the umbilical cord, without any apparent intrinsic value – reconfigure the relationships among the various actors involved (Waldby 2002). All these actors, be they researchers, physicians, patients, public institutions and private enterprises, donors, recipients, caregivers, or others besides, have citizenship rights/duties which, in their turn, are necessarily redefined in the process of biovalue creation. Citizenship of a society which envisages the use of biobanks thus assumes a different meaning from that in the past. In other words, the creation and operation of biobanks for scientific research entails a partial reformulation of that meaning.

On the other hand, the contradictory aspects apparently inherent in biobanks depend on the nature of that particular form of social relationship which we call *gift-giving*. As Mauss clarified, giving does not consist solely in the transfer of a good from one person to another; it also necessarily entails that the donor transmits a part of him/herself to the other. Otherwise, it is not a gift that is made, but a loan. Nevertheless, again according to Mauss, this does not rule out that the gift-giver does not expect a future quid pro quo: giving requires reciprocation, a commitment to solidarity and the social order which, however, is honoured at a distance of time (Mauss 1923–1924).

The social logic of the gift described by Mauss is evidently at work in the archetype of all biobanks, namely the blood bank (Busby 2006). In fact, already inherent in the birth and institutional development of blood donation was a duality again to be found in the debate on the regulation of biobanks. Once blood has been extracted from the body and inserted into the context of scientific research and medicine, it acquires a dual character whereby it constantly oscillates between being a public good that is donated and being a valuable resource that can be exploited for commercial ends. Donated blood can be defined as simultaneously a collective good extraneous to the logic of commercial exchange and as an individual good which is instead subject to that logic. But also as a public good, donated blood is ambiguous: the appeal to altruism made by blood donation campaigns, in fact, contains an implicit prospect of individual future benefit, both because one day the donor or one of his or her loved ones may need a transfusion, and because the medical treatments deriving from blood-based research may sooner or later prove useful. Thus a donation made today creates expectations of a future return. As it has been noticed for the case of cord blood.

> in both temporal and spatial terms, cord blood storage, like other forms of biobanking, spans time by projecting the investment of bodily tissues into a future where the potential value of that investment "might" be redeemed, (Brown 2005 345)

what Thompson calls the *promissory* characteristic of capital in biomedical enterprise (2005: 258–60). Conversely, the potential individual benefit reinforces the solidaristic motivation, so that the two dimensions are inextricably bound up with each other.

Hence,

> the dual nature of blood provides a certain elasticity to the deployment of blood and blood products in public discourse. (Busby 2006: 854)

This creates an important precedent for biobanks, given that

> policy advocates for development of biobanks find it quite easy to switch registers in their framings – here it is a public good and then in a flash it is a resource to be exploited [. . .] – blood in political discourse is both priceless and valuable. (Busby 2006: 854)

It is interesting to note that the "gift" metaphor in regard to biobanks works as a discursive strategy with which to resolve the contradiction due to the twofold nature of the organic materials that they collect and store. In fact,

> the gift language of professional guidelines formalizes an ethical discourse, incorporating while delimiting property rights for the use of tissue

and introducing a further distinction

> between tissues themselves and the information derived from them. (Busby 2006: 859)

It is this distinction that makes it possible to separate

> two economies, that of non-commercial tissue donation complete with a discourse of altruism, and the commercial realm of genetic information that can be transformed into property. (Tutton 2004: 33)

Moreover, we may extend to biobanks in general what others have observed regarding the public discourse on specific types of biobank. On the one hand, one finds

> crucial differences in emphasis and strategy between public and private cordon blood (CB) banks: while the commercial sector cites yet-unrealized future developments in tissue engineering as a primary reason for investing in CB banking, the public sector focuses almost exclusively on the present-day use of CB human stem-cells in treating very rare blood and immunological disorders, as well as in rare instances where a bone marrow transplantation is not possible. (Brown 2005: 341–2)

On the other hand, one should not overlook the fact that

> the appeal to altruism, however, when combined with the metaphor of donated blood as "gifted", has come to obscure the choices that are to be made about the organisation and boundaries of a public genetic research biobank. The nostalgic cast of this metaphor has

> obscured discussion about the research uses of diverse collections of genetic material held by public bodies in the UK, and has limited the scope of formative policy discussions about oversight of the new national biobank. (Busby 2006: 861)

Finally, what we think, say, and do about biobanks would not be possible without the widespread and socially rooted conception of health which incorporates the idea of risk; a conception which, in its turn, is supported and reinforced by the public discourse on biobanks. This is a notion of health very different from traditional ones, in that it is widened until it redefines all citizens as potentially ill and, therefore, as legitimate objects for scientific research.

Many authors have emphasized the role of biobanks in opening new health markets based on risk. Biobanks are accordingly part of a more general process of redefining the conception of health, which tends increasingly to merge with the idea of risk. A biobank operates on a logic with features entirely consistent with this tendency: (a) it is prospective, in that it gathers clinical and environmental data before subjects manifest disorders, and not afterwards; (b) it works as a platform which, rather than being concerned with individual pathologies, is designed to be multipurpose; (c) it is cumulative, in the sense that each participant enriches the overall sample; (d) it has particularly broad dimensions because it works on genetic differences, environmental factors, and the presence of disorders, a characteristic which once again favours the encounter with the public health service, as an agency which standardizes and harmonizes the numerous differences among individual biobanks (Mitchell and Waldby 2009: 12–4).

The encounter between the increasing concern of Western societies with bodily care and post-genomic diagnosis tools is a further factor fostering the growth of a conception of health imbued with the risk dimension. Whereas classic genetics sought the genetic causes of pathologies, post-genomic genetics work with not a causal but a probabilistic logic, in that they include people who are well and would traditionally have been considered healthy in the category of the sick precisely because they are exposed to the risk of becoming such.

5.3 Body and Identity

A conception of health closely bound up with the idea of risk reverses the issue of the relation between body and identity in the current social context; a relation of prime importance in the formation of bio-value through biobanks and the gift-giving dynamic involved in it.

Although the body may no longer be the biblical "temple of the soul", it certainly appears to be the seat of the identity (Nelkin and Lindee 1995; van Dijck 1998). It is through the mediation of the body that the *DNA* = *identity* equation becomes plausible; the cultural resilience of this equivalence is explained by the revaluation of the body as the seat of identity: DNA is the body, and the body is identity.

Not by chance, the importance of the body in the construction and maintenance of identity has been emphasized as a key feature of post-modern society. The body functions as an increasingly indispensable material support and a symbolic referent as the traditional bases of identity are eroded by the reflexive application of the founding principles of modernity or, at any rate, by the disappearance of the certainties of modernity (Beck 1992; Giddens 1991; Bauman 1995; Featherstone et al. 1991; Shilling 1993). We increasingly cling to the idea that "we are our bodies", because the body seems to be the last bulwark left after the dissolving of other social and cultural certainties concerning our identity (Giddens 1990; Melucci 1996).

The body performs a crucial function in this new dimension as the "receiver of sensations" (Bauman 1995: 111–3). Hence, if the identity is founded on a constant search for new experiences, the body becomes the material and symbolic referent in which to anchor the self. The body figures in our culture as both the material support for our existence and the icon of our identity. It is therefore a good to be cherished, exhibited and conserved (Featherstone et al. 1991; Giddens 1990, 1991; Shilling 1993; Melucci 1996; Bauman 1995; Lash 1979).

The body's centrality in our society is also expressed in the fact that we mainly talk about the body in individualistic rather than collective terms.

This tendency is encouraged by the fact that

> the increasing specificity of diagnosis matched by ever more targeted tests appears to make medicines more oriented to individuals.

Of course,

> on the other hand, informational technologies enable data to become more abstracted at the level of population. (Hogle 2008: 848)

But even in the specific case of biobanks, the shift to the collective level is never complete nor definitive, especially in the case of initiatives involving commercial interests. The potential impact on people's health is a constant referent, and the central importance of the body of the single individual is never seriously disputed.

But this centrality of the body corresponds to its fragility. High modernity has created a context in which identity becomes a "reflexively organized project": a construction as crucial as it is fragile, and which gives great importance to the body, the principal foundation of a highly unstable identity and the best way to give it tangible form (Giddens 1991). Also thanks to science, our body has become more opaque. For example, we know a lot about our bodies, but we do not know when life begins or when it ends.

This gives rise to a profound contradiction: whilst the body is an increasingly crucial referent for the identity, the conditions that attribute it that role undermine socially shared conceptions of what the body is, how it should be constructed or protected, and about the relationship between body and identity. We thus become dramatically aware that

> where biology and nature end and culture begins is not decided outside culture. This is the new paradox. We are thus overcultural beings facing the necessity of deciding on our own nature. (Melucci 1997: 69)

Amongst other things, the fragile as well as crucial nature of the relationship between body and identity in contemporary societies acts as cultural foundation which legitimates biobanks and fuels the social practices necessary for them to function. It is in this context that biobanks can be viewed as advance deposits against the corruption of the body by illness, as insurance on our future health entrusted to biomedical research. Moreover, this applies at both the individual and collective level. It thus favours resolution of the contradiction apparently inherent in the practice of gift-giving: making part of one's body available to biomedical research protects a fragile and valuable good, and it in the meanwhile strengthens the social bond through the exercise of solidarity, even if it is not so disinterested as it could be believe at a first glance.

5.4 Conclusions

Biobanks unequivocally lie at the intersection of a series of processes which gainsay any idea of a separation between science and society. At the same time, precisely these processes have progressively created the premises for biobanks to become part of our social landscape.

Viewing biobanks from the perspective of the co-production of science and society also makes it possible to move beyond the contradiction – sociologically only apparent – which opposes public utility and private profit, gratuitousness of the gift made on behalf of the collective good and individual profit in the form of economic remuneration or the safeguarding of one's health, through an advance deposit of biovalue as insurance against illness.

It is likewise clear how little reliance can be placed in distinctions now superseded by the facts to govern the array of social practices entailed by biobanks. Consider, for example, the separation between pure and applied research. We cannot but acknowledge that its disappearance further complicates matters concerning biobanks, and that some nostalgia for this distinction seems justified. Nevertheless, attempts to separate them artificially, for example by means of rules intended to distinguish between the use of materials conserved in biobanks for the purposes of pure research (made to coincide with public research) and their use for applied research (typically that conducted in the private sector), or through the rhetorical use of such a distinction in discourses on biobanks, does not seem an appropriate strategy.

This is firstly because the interweaving between the theoretical and practical dimensions has always been part of scientific progress. It is therefore a phenomenon which is by no means new, for it has been intrinsic to the development of science and technology from their beginnings. As it is well-known, for Bacon the meaning of the quest for knowledge which began with nascent modern science lay in its applicational value, helping mankind to bend the laws of nature to its purposes: *scientia potentia est.* Subsequently, awareness of the nexus between knowing and doing became even more refined, until Vico synthesised it into the dictum *verum est factum*: we can only fully understand that which we are able to do, to reconstruct.

It is secondly because developments in genetic engineering, neuroscience and nanotechnology, along with developments in the debates that accompany them, has taken this interweaving to its extreme consequences, not so much because of its presumed impact on society or what are generally defined as the social consequences of scientific progress, but rather because

> it is necessary to produce offspring in test tubes, it is necessary to create genetically artificial beings, it is necessary to build reactors, before and so that their properties and safety issues can be studied. (Beck 1992: 61)

In other words, the increasingly close interpenetration between science and society makes it unthinkable to postpone evaluation of the effects of scientific knowledge to after its production, because the former are manifest as the latter proceeds. We will know the effects of research conducted with biobanks when they already manifest in the social fabric, but at that point they will already be an part of it.

Biobanks, as the crowded crossroads of social processes which they also help drive, and thanks to which they are able to exist, must inevitably be hybrid in nature.

On one hand, the *pieces of ourselves* which are conserved in biobanks must be made independent from our bodies, not just materially but also, and especially, symbolically: only in this way can they become resources available for producing biovalue, for acquiring value as a gift to scientific research (which transforms them into a collective good by transforming them into scientific knowledge).

On the other hand, it is this very gift which implies that the donor's detachment is never complete, not even on a symbolic level: the organic material conserved in a biobank is always a part of someone; it is always a part of their identity. Only this is a real gift. Besides, the value (whether scientific, commercial, financial or political) of the information which can be extracted from the organic material conserved in biobanks depends on the maintenance of a link with its original owner (who was afflicted with a certain illness, the owner of a certain individual and family history, lived in a particular context and in a given period of history). The organic matter conserved in biobanks acquires value when it can be traced to the person to whom the body belonged.

Biobanks, furthermore, produce scientific knowledge useful for curing a body, that precious and fragile material support for individual identity. It is this socially-shared conviction which sustains the construction of biobanks, and which, at the same time, makes the necessary investment and the prospect of future commercial gain plausible.

Biobanks are certainly part of our society because of the intrinsic value which is socially granted to scientific knowledge, especially to that produced by biomedical research, and because of their expected commercial value; this is a value, however, that is due to the importance of biobanks for biomedical research.

Individual and collective investment in biobanks for scientific research is therefore configured in terms of "scientific imagery" (Fujimura 2003), and it is not directed solely to oneself or loved ones. It is also directed towards a future collective gain definable as an expected increase in socially available cures. At the same time, the credibility of considering as a possible future gain what biomedical research can demonstrate only in terms of "scientific imagery" is based on the increasingly widespread social desire for bodies which are always efficient, healthy bodies which guarantee the autonomy of individuals and anchor them to something which seems solid but is instead uncertain.

References

Bauman, Z. 1995. *Life in Fragments: Essays in Postmodern Morality*. Oxford: Blackwell.

Beck, U. 1992. *Risk Society: Toward a New Modernity*. London: Sage.

Brown, N. 2005. "Shifting Tenses: Reconnecting Regimes of Truth and Hope." *Configurations* 13: 331–55.

Busby, H. 2006. "Biobanks, Bioethics and Concepts of Donated Blood in the UK." *Sociology of Health & Illness* 28 (6): 850–65.

Featherstone, M., M. Hepworth, and B. Turner. 1991. *The Body: Social Process and Cultural Theory*. London: Sage.

Fujimura, J. 2003. "Future Imaginaries: Genome Scientists as Sociocultural Entrepreneurs." In *Genetica Nature/Culture: Anthropology and Science Beyond the Two-Culture Divide*, edited by A. Goodman et al. Berkeley: University of California Press.

Giddens, A. 1990. *The Consequences of Modernity*. Cambridge: Polity Press.

Giddens, A. 1991. *Modernity and Self-Identity*. Stanford: Stanford University Press.

Haddow, G., L. Graeme, S. Cunningham-Burley et al. 2007. "Tackling Community Concerns About Commercialisation and Genetic Research: A Modest Interdisciplinary Proposal." *Social Science & Medicine* 64: 272–82.

Hogle, L.. 2008. "Emerging Medical Technologies." In *Handbook of Science and Technology Studies*, edited by E. Hackett et al. Boston: MIT Press.

Jasanoff S. (Ed.). 2004. *States of Knowledge: The Co-production on Science and Social Order*. London: Routledge.

Keating, P., and A. Cambrosio. 2003. *Medical Platform: Realigning the Normal and the Pathological in Late-Twentieth-Century Medicine*. Boston: MIT Press.

Knorr-Cetina, K. 1981. *The Manufacture of Knowledge: An Essay on the Constructivist and Contextual Nature of Science*. Oxford: Pergamon.

Lash, C. 1979. *The Culture of Narcissism*. New York: Norton & Co.

Latour, B. 1984. *Les microbes: guerre et paix, suivi de Irreductions*. Paris: Metailie.

Latour, B. 1987. *Science in Action: How to Follow Scientists and Engineers Through Society*. Harvard: Harvard University Press.

Latour, B. 2005. *Reassembling the Social: An Introduction to Actor-Network-Theory*. Oxford: Clarendon.

Latour, B., and S. Woolgar. 1979. *Laboratory Life*. Princeton: Princeton University Press.

Mauss, M. 1923–1924. "Essai sur le don. Forme et raison de l'échange dans les sociétés archaïques". Année Sociologique (Seconde série).

Melucci, A. 1996. *The Playing Self: Person and Meaning in the Planetary Society*. Cambridge: Cambridge University Press.

Melucci, A. 1997. "The Social Production of Nature." In *Gene Technology and the Public*, edited by S. Lundin and M. Ideland, 58–70. Lund: Nordic Academic Press.

Mitchell, R., and C. Waldby. 2009. "National Biobanks: Clinical Labour, Risk Production, and the Creation of Biovalue." *Science, Technology and Human Values*. doi: 10.1177/0162243909340267.

Mulkay, M. 1979. *Science and the Sociology of Knowledge*. London & Boston: Allen & Unwin.

Nelkin, D., and S. Lindee. 1995. *The DNA Mystique: The Gene as a Cultural Icon*. New York: Freeman.

Palsson, G., and P. Rabinow. 1999. "Iceland: The Case of a National Human Genome Project." *Anthropology Today* 15 (5): 14–18.

Palsson, G., and P. Rabinow. 2001. "The Icelandic Genome Debates." *Trends in Biotechnology* 19 (5): 166–71.

Shilling, C. 1993. *The Body and the Social Theory*. London: Sage.

Sigurdsson, S. 2001. "Yin-Yang Genetics, or the HSD Decode Controversy." *New Genetics and Society* 20 (2): 103–17.

Testa, G. 2006. "Che cos'è un clone? Pratiche e significato delle biotecnologie rosse in un mondo globale." In *Cellule e cittadini*, edited by M. Bucchi and F. Neresini. Milano: Sironi.

Thompson, C. 2005. *Making Parents: The Ontological Choreography of Reproductive Technologies*. Cambridge: MIT Press.

Tutton, R. 2004. "Person, Property and Gift: Exploring Languages of Tissue Donation to Biomedical Research." In *Genetic Databases: Socio-Ethical Issues in the Collection and Use of DNA*, edited by R. Tutton and O. Corrigan. London: Routledge.

van Dijck, J. 1998. *ImagEnation: Popular Images of Genetics*. New York: New York University Press.

Waldby, C. 2002. "Stem Cells, Tissue Cultures and the Production of Biovalue Health." *An Interdisciplinary Journal for the Social Study of Health, Illness and Medicine* 6 (3): 305–23.

Part II
The Rights of Donors and Patients

Chapter 6
One Sample, One Share! A Proposal to Redress an Inequity with Equity

Jasper Adriaan Bovenberg

6.1 Prologue

On September 17, 2010, Johnson & Johnson (NYSE: JNJ) and Crucell N.V. (NYSE Euronext, NASDAQ: CRXL; Swiss Exchange: CRX) announced that they were in advanced negotiations for a potential public offer for all outstanding ordinary shares of Crucell not already held by JNJ, for approximately €1.75 billion, which represented a purchase price of €24.75 per share. According to the press release announcing the bid, JNJ intends to continue to invest in the development of Crucell's products and pipeline. The main driver behind these products and pipeline is the PER.C6® cell line, an immortalised factory for the manufacturing of pharmaceuticals and vaccines. This cell line is derived from a single, human retina-derived cell, which was purposely immortalised using recombinant DNA-technology as follows:

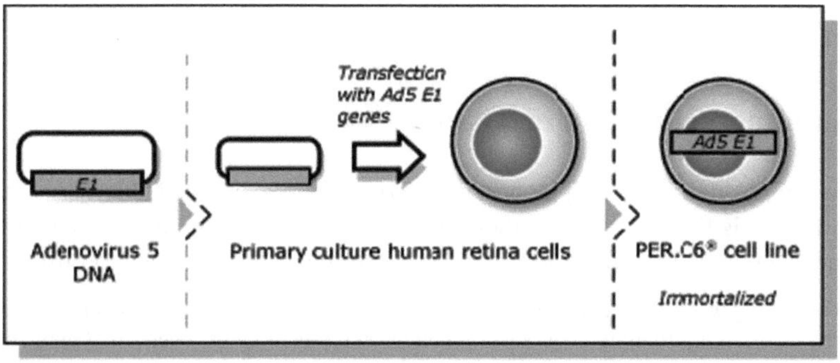

Fig. 6.1 Crucell's E1 cell line technology (source: http://www.crucell.com/Description)

Crucell maintains extensive documentation on the origin, establishment, and characterization of the PER.C6® cell line, establishing it as the most completely

J.A. Bovenberg (✉)
Legal Pathways Institute for Health and Biolaw, Aerdenhout, The Netherlands
e-mail: jabovenberg@xs4all.nl

C. Lenk et al. (eds.), *Biobanks and Tissue Research*, The International Library of Ethics, Law and Technology 8, DOI 10.1007/978-94-007-1673-5_6,

documented cell line to date. The person who donated the cell, however, is not known to Crucell. Most certainly, it is a cell from a foetus which was donated for scientific research. This "donation" took place at a time when (maternal) consent was not required.

6.2 Introduction

The Crucell story is but one of many illustrations that the biomedical research enterprise is built on the contributions of the trinity of donors (*samples*), universities (*know why*) and industry (*know how and capital*).[1] It is also one of multiple illustrations of the double standard which continues to govern the commercialization of biological materials. While the law allows universities and industry to capitalize on their contributions, it denies donors of biological materials *both* the right to compensation for *and* the right to control the use of their contributions.[2] To resolve this donor "cash and control deficit", many solutions have been advanced, ranging from direct control through the recognition of inalienable property rights in the tissue,[3] to indirect forms of control (co-determination by donors through governance mechanisms[4]) and forms of benefit sharing through a tissue-tax.[5] This paper explores a novel solution. Both universities and industry capitalize on their contributions by contributing them to a corporation in exchange for shares in the corporation's capital. This form of capitalisation triggers an obvious, but hitherto unasked question: if inventors and investors can contribute in exchange for shares, then why can't donors? Could issuing shares in exchange for samples redress the "gain and governance" deficit currently felt by sample donors? This paper will examine the pros and cons of this "shares for sharing" model.

6.3 Commercialisation of Tissue

Rightly or wrongly, a major concern, if not a source of deep distrust, for many tissue donors is the potential of their sample being commercialized by a (foreign) company.[6] As observed in a report by the Australian Law Reform Commission:

> Another thing that clearly emerged at the public forums is the atavistic or primal fear among members of the community about their genetic material being sent "overseas" (again, often expressed as being "sent to the US"). So, at almost every event, someone in the audience expressed concern about volunteering for an experiment at an Australian university research

[1] For example the story behind the HeLa cell line, in Skloot (2010).
[2] Bovenberg (2005).
[3] Laurie (2002, 318); Gitter (2004, 257).
[4] Winickoff and Winickoff (2003, 349, 1181).
[5] Bovenberg, supra note 3.
[6] Kanellopoulou (2011).

> lab or teaching hospital, then finding that the research group had "spun off" into a private biotech company, which then merged with or was taken over by American interests — and "the next thing you know, your DNA is overseas"![7]

Indeed, concerns over commercial use of freely donated tissue has plagued the (funding of) private and public-private efforts to set up national biobanking initiatives in Iceland and Estonia.[8] Surveys in European countries suggest commercialisation of freely donated tissue (and data) is among the highest concerns among the European publics as regards biobanking.[9] As a high profile scientist/corporate insider has put it,

> the single biggest thing that keeps this from happening is people being concerned that someone else is going to get rich in an unfair way and they're not.[10]

Anxiety over commercialisation is problematic as the same publics that oppose commercialisation do demand that research produces tangible results in the form of clinical and health outcomes, i.e. therapies and drugs. The best way to get these outcomes, however, still seems to be allowing commercial parties to transform samples and knowledge into marketable products, technologies and platforms. As there is then, a good case for allowing the commercialisation of tissue, the question is why the donors are not allowed to govern and share the proceeds.

6.4 Proposed Solutions

As we saw in the introduction, to redress this donor "gain and governance inequity", many proposals have been advanced. Some suggest that donors should leverage their (collective) rights to consent or not to consent to the use of their tissue by negotiating favourable terms with academia and industry for the use of their tissue.[11] This model has been successfully applied by such patient advocacy groups as PXE International. Others have proposed to recognise inalienable property rights in their tissue, which rights could then also be used as bargaining power to exact proper quid(s) pro quo from academia and/or industry for the use of their tissue.[12] While these solutions have turned out to work in some cases, for most cases they may not work for reasons of creating an anti-commons.[13] Also, they will require lengthy negotiations and result in contractual rights vis-à-vis academia and industry concerned. The contract model, however, does not provide for institutional engagement of donors

---

[7] Australian Law Reform Commission (2004).

[8] Greely (2000); Potts (2002).

[9] For example various public consultations for UK Biobank. Available at: http://www.ukbiobank.ac.uk/docs/perceptions.pdf (accessed 07 March 2011).

[10] Dr David Cox, senior vice-president at Pfizer, in *BBMRI: the Industry Perspective*. Available at: http://www.pharmaceutical-technology.com/features/feature97506/ (accessed 07 March 2011).

[11] Terry (2003, 377–93).

[12] Laurie, supra note 3.

[13] Bovenberg (2006), chapter 6.

at the company level and is prone to overstretch the actual value of the contribution of one or more samples to the eventual success of a biomedical research enterprise.

If we view the collaborative research enterprise model for what it actually is, then another solution emerges. Commercial use of human tissue is brought about by the integration of the three classic ingredients of free enterprise: labour, capital and raw material. If the value of the contributions of both labour and capital can be expressed in the form of (employee options on) shares (stock) in the enterprise's capital, then the contribution of raw material (samples) might also be so appreciated. Indeed, issuing shares for samples would unveil new perspectives, which would do justice and give real meaning and power to enforce a range of donor rights, commensurate to the value of the contribution. To test this hypothesis, it will be applied to a typical form of corporation: the Dutch closed corporation with limited liability or "b.v.", aka in Germany as the "Gesellschaft mit beschränkter Haftung" (GmbH) and in France as the "societé à responsabilité limitée (s.a.r.l.)".

6.5 Closed Corporation with Limited Liability

The Dutch closed corporation with limited liability is a legal entity with an "authorized capital" divided in shares. Share certificates are not being issued and the shares are not freely transferable. A shareholder is not personally liable for any liabilities incurred by the corporation and is not obliged to share in the losses of the corporation in excess of the amount of his contribution.[14] Shares are the parts in which the authorized capital has been divided, e.g. an authorized capital of €200.000.= divided in 2000 shares of €100 .=.[15] The authorized capital indicates the maximum amount for which shares can be issued. A share must be expressed in a nominal value and when a share is issued, its nominal amount must be paid up.[16]

Paying up for a share must take place in cash, unless another form of contribution has been agreed upon: the so-called *contribution in kind*.[17] A contribution in kind requires that the contribution can be assessed in economic terms.[18] The incorporators must make a description of the contribution, including the value attached to it and the valuation methods used to calculate this value. This description must be accompanied by a declaration of an accountant to the effect that the value of the contribution is, using generally accepted accountancy methodologies, at least equal to the amount that must be paid up on the share.[19]

[14] Art. 2:175 Dutch Civil Code.

[15] Art. 2:195 Dutch Civil Code.

[16] Art. 2:191 Dutch Civil Code.

[17] Art. 2:191a Dutch Civil Code.

[18] Art. 2:191b Dutch Civil Code.

[19] Art. 204a Dutch Civil Code. For contribution in kind after the incorporation, art.204b Dutch Civil Code provides for a similar mechanism.

6.6 Human Tissue as a Contribution in Kind?

A contribution in kind implies that the contributed "kind" is transferred from the contributor to the corporation. Upon this contribution, the good will form an asset of the corporation to which it has been contributed. This raises the question of whether donors can transfer their rights in their tissue. Under Dutch law, natural persons can "dispose" of their "goods". A good has been defined in Dutch law as "any material capable of human control".[20] As the Crucell case and many others, demonstrate, human tissue is indeed "capable of human control", even up to the point of being immortalized.

While there is no specific provision in Dutch law to that effect, it is generally held that a human being is the owner of his tissue. The next question then is whether a human being, being the owner of his tissue, can transfer this ownership (by way of making a contribution in kind) to a third party (i.c. the corporation). Ownership rights are, in principle, capable of being transferred, unless the law or the nature of the right opposes such transfer. There is no Dutch law which opposes the transfer of ownership of human tissue. There are, indeed, a number of laws that prohibit the offer and acceptance of monetary consideration for donation of certain types of tissue, such as fetal tissue, blood, or organs. These prohibitions as such, however, do not rule out that the ownership of these and other tissue be transferred; they only rule out that the donor is paid for a transfer. The laws do not prohibit the subsequent use of such tissue for commercial purposes; in fact they promote such use. The eligibility of tissue for transfer can also be derived from the fact that said laws do allow for the donation of the tissue. Such donation implies transfer of ownership, even if this ownership is limited in the sense that it cannot be transferred in exchange for financial consideration.

A related issue is whether, by contributing his sample to the corporation, the sample shareholder has also transferred his personal rights to control the use of his tissue. Can he still object to certain types of uses? Can he still withdraw from research on his sample, at will? It could be argued that, while the ownership of the sample will have been transferred as a result of the contribution in kind, certain personality rights have not, if and to the extent that they must be considered "inalienable". This interpretation, however, sits uncomfortably with the notion, and the expressed desire, of the sample contributor to share in the proceeds of the use of his sample: "you can't have your cake and eat it too". While contribution of his sample may, at least in theory, act to deprive a sample contributor from the opportunity to vote with his feet against uses of the sample he finds objectionable, use of the sample is by no means unregulated. Statutory and regulatory requirements prohibiting or restricting unethical use of samples continue to apply, as do medical review board review and supervision of sample research.

---

[20] Art. 2:3 Dutch Civil Code.

6.7 Is a Share a Financial Gain?

A related question then, is whether the contribution of tissue in exchange for the issue of a share in the capital of a corporation to which the tissue is contributed, qualifies as a contribution in exchange for a financial consideration? Such a financial consideration, as we have seen, would be prohibited for at least certain types of tissue and, under the Oviedo Convention, for all types of body parts.[21] Prima facie, a share in the capital of a corporation represents an economic value, as is evidenced by the very requirement that the contribution in kind for such a share be assessable in economic terms and must be valued at least the nominal amount of the share. A share as such, however, does not necessarily represent a "financial gain". It may be a share in a corporation with negative equity; the corporation may be a loss-making entity from the start and it may never make a profit. In fact, many a biotech start-up using human tissue never makes a dime, let alone ever pays any dividends to its shareholders.

Clearly, shares in a loss making corporation do not represent a "financial gain". So, while the value of a share may be positive (e.g. JNJ offered Crucell €24.75 per share), it may be very well be negative. The fact that the tissue (must) have an economic value does not render the issue of a share a "financial gain". In fact, the rationale of contributing the tissue to the corporation is to contribute to a research enterprise whose very aim it is to find out *whether* any financial gain can be made, which enterprise may succeed, or not. If anything, then, a share represents a financial risk, rather than a financial gain.

6.8 Shareholders' Rights: To Gains and Governance

Having been issued, in exchange for the contribution of his tissue, a share in the share-capital of the corporation, the sample donor has become a shareholder in the share-capital of the corporation. As a shareholder, he enjoys shareholder's rights. These rights can, roughly, be distinguished in "control" rights or "governance" rights and "capital" or "gain" rights. The "governance" rights are bestowed to the general meeting of shareholders. This corporate body has all the powers, within the limits set by the law and the articles of association, which have not been bestowed on the board of directors or others.[22] Both the board of directors and the supervisory board, if any, must provide the general meeting of shareholders with all requested information, unless in the event of a serious interest of the corporation to the contrary.[23] This right enables an individual shareholder, such as the tissue donor, to

---

[21] Article 21 Convention for the protection of human rights and dignity of the human being with regard to the application of biology and medicine: Convention on Human Rights and Biomedicine, Explanatory Report (ETS No. 164), no. 133.

[22] Art. 2:217 (1) Dutch Civil Code.

[23] Art. 2:217 (2) Dutch Civil Code.

be provided with all requested information.[24] Thus, this shareholders' right is a valuable tool for the tissue donor to be informed of what the corporation is doing with his tissue and thus could help satisfy a fundamental right of tissue donors to be informed. Indeed, there probably is not a single academic, public clinical or population biobank consent based governance system which grants its donors such extensive rights.

The right to information about the use of the tissue might be even better guaranteed in the event the corporation would not be a closed corporation but a public corporation listed on the stock exchange. The disclosure requirements for listed corporations will result in extensive information being provided to the shareholders. The 2009 Crucell prospectus, for example, contains detailed information on the commercial whereabouts of the *PER.C6® cell line* as shown in its Table of Contents:

Table of Contents

Chapter	Page
1 Summary.	2
2 Risk Factors.	8
3 Important Information.	18
4 Dividend Policy.	23
5 Use of Proceeds.	24
6 Capitalization and Indebtedness.	25
7 Selected Financial Data.	27
8 Operating and Financial Review.	29
9 Industry Overview.	49
10 Business Overview.	54
11 Management Board, Supervisory Board and Employees.	81
12 Major Shareholders.	99
13 Related Party Transactions.	102
14 Share Capital and Corporate Governance.	104
15 Taxation.	114
16 General Information.	117
17 Definitions.	119

Fig. 6.2 Providing transparency, if not full disclosure, on sample use: the Table of Contents of the Crucell NV 28 October 2009 prospectus re the "Admission to trading of 14,626,984 newly issued ordinary shares with a nominal value of €0.24 per share"

The "governance" rights further give the general meeting of shareholders the following powers: the power to change the articles of association[25] (including the purpose of the corporation), dissolution of the corporation,[26] mergers with other corporations,[27] conversion of the corporation into another legal entity,[28] and dilution.[29]

[24]Some commentators question, however, whether this right of the general meeting of shareholders extends automatically to a single shareholder or group of shareholders.

[25]Art. 2:231 Dutch Civil Code.

[26]Art. 2:19 Dutch Civil Code.

[27]Art. 2:317 Dutch Civil Code.

[28]Art. 2:18 Dutch Civil Code.

[29]Ibid.

In addition, the general meeting of shareholders has the power to appoint, suspends and fire the board of directors, at least two-thirds of the supervisory directors[30] and to determine the annual accounts.[31] These powers of the general meeting of shareholders enable the shareholders to govern the corporation. To the extent the corporation is dealing with human tissue and to the extent the shareholders are tissue donors, they can exercise their corporate powers, commensurate, of course to the proportion of their shareholding, to ensure that the use of their tissue is in their best interest. Again, there probably is not a single academic, public clinical or population biobank consent based governance system which grants its donors such extensive rights.

The above powers lie with the general meeting of shareholders. In addition, the law also bestows certain rights on the individual shareholder. He or she may call for a meeting of the shareholders, should the board of directors or the supervisory board fail to do so. Every single shareholder is entitled to attend the general meeting of shareholders, to present his opinion and to exercise his voting rights.

6.9 One Share, One Vote

The central principle of the law is that the shareholders have powers that are proportionate to their shareholding. Only shareholders have voting rights and every shareholder has at least one vote.[32] While the power within the corporation lies, eventually, with the general meeting of shareholders, that does not guarantee that the power is exercised in the exclusive interest of the tissue-shareholder. In practice, tissue-shareholders are likely to be minority shareholders, which can be outvoted by the majority shareholders. The primary opportunity for (minority) tissue-shareholders to ensure that their tissue is used in their best interest would be for them to seek to establish a majority vote. Even absent a majority shareholding, however, sample-shareholders could prevail. First, the law allows for two exceptions to the rule that the voting right is proportionate to the nominal amount of the share. The articles of association may limit the number of votes per shareholder, under certain conditions. The article may also deviate in another way from the proportionality rule, provided the voting right per shareholder is limited to six, respectively three votes.

6.10 Priority Share

Second, the law allows for the issue of special shares. One such category of special shares are the so-called "priority shares". Priority shares are typically created to subject certain decisions to the approval of the meeting of priority shareholders.

[30] Art. 2:242 and 2:244 Dutch Civil Code.

[31] Art. 2:210 Dutch Civil Code.

[32] Art. 2:228 Dutch Civil Code.

The category of priority share decisions may be both management decisions (e.g. the decision to re-finance, to make substantial investments or to pay an interim-dividend) and decisions which are the prerogatives of the general meeting of shareholders (e.g. the issue of new shares, payment of dividend). By requiring its approval, priority shareholders are effectively given a veto on those decisions. One specific example of a right to be given to the priority shareholding is the right to set the number of directors and the right to make a binding proposal for the appointment of (a certain number of) directors. Overruling such a binding proposal requires a two-third majority in the general meeting of shareholders in which more than half of the issued share capital is represented.[33] Obviously, specific priority shares can be designed so as to subject pertinent decisions regarding the use of sample(s) to the approval of the pertinent sample-priority shareholder(s).

6.11 Shareholders' Capital Rights

In addition to (shared) governance powers, shareholders have rights to the capital. The reward for a shareholder for his contribution to the share capital of the corporation may be a dividend, an increase of the value of his share (capital gain) or a combination thereof. As a shareholder, then, a tissue donor will be entitled to the dividend and any capital gains resulting from the corporation's use of his tissue. In the case of Crucell, a sample-shareholder would have experienced the following share performance:

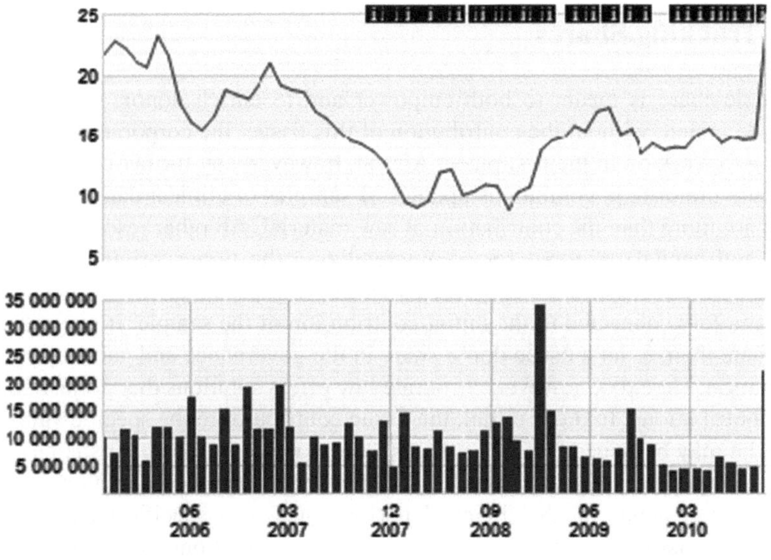

Fig. 6.3 Crucell share performance (source: Thomson Reuters. The information so distributed is not intended for operative purposes. Reuters suggests that you consult your financial intermediary before any operation)

[33] Art. 2:243 Dutch Civil Code.

The dividends are paid to the shareholders in proportion to their shareholding. It is possible, however, to create, in the articles of association, different kinds of shares, which each have their own right to dividends. The different kinds of shares are typically indicated by a different letter, hence their name "letter shares". "Letter shares" provide for the opportunity to pay out dividend to one group of shareholders, while reserving dividend for another group of shareholders. Dividends are profits of the corporation, as apparent from the determined annual account of the corporation. The shareholders can decide to pay out a dividend or make a whole or partial reservation of the profits. Whether a dividend will actually be declared is the prerogative of the general meeting of shareholders. Crucell, for example, has not paid any dividends during the last three financial years, and it intends to retain future earnings, if any, to finance the growth and development of its business.[34]

Regardless of type of share, a sample-share holder will be entitled to dividend if such a dividend is distributed. Thus, he or she will be able to share in any profits the corporation may make. Thus, being a shareholder in the corporation that is commercializing his or her tissue would be the ultimate form of benefit-sharing. Notably, a dividend can only be declared if the corporation, applying all contributions made to it by the various contributors, has actually made a profit on its use of the contributed tissue. Admittedly, the sample-shareholder will only be entitled to a dividend that is proportional to his shareholding. This, however, can hardly be seen as unjust. His contribution of his tissue was only one fraction of the resources (inventions and investments) that were needed to make the corporation succeed.

6.12 Tracking Shares

A sample-share is meant to both empower and to enrich donors of samples. Its rationale is that, without the contribution of this tissue, the corporation would have nothing to work on in the first place. Absent, however, any further contributions by the tissue-donor (e.g. continuous updates of samples or clinical data), his contribution is not more than the contribution of raw material. All other resources (intellect, skills, and capital) required for commercializing the tissue, originate from other sources and the corporation may very well develop products and services that are only remotely connected to the initial contribution of the sample. It may be disproportionate then, to let a tissue donor share in the governance and gains of the *entire* corporation. Here too, however, corporate law offers solutions that do justice to relative contributions. In order to link the tissue contributor to the specific value of his tissue, he may be issued "tracking shares", also known as "designer shares".

As their name suggests, "tracking shares" are a type of common stock that "tracks" or depends on the financial performance of a specific business unit or operating division of a company—rather than the operations of the company as a

[34]Crucell, Prospectus dated 28 October 2009, Dividend Policy, p.23.

whole.[35] When a parent company issues a tracking share, all revenues and expenses of the applicable division are separated from the parent company's financial statements and bound to the tracking share. Tracking shares trade as separate securities. As a result, if the unit or division does well, the value of the tracking shares may increase – even if the company as a whole performs poorly. The opposite may also be true. Shareholders of tracking shares have a financial interest only in that unit or division of the company.[36]

Applying the concept of "tracking shares" to sample-shares, the sample-shares could be designed so as to track or depend on the financial performance of the unit of the corporation that holds the tissue-sample or that directly capitalizes on the tissue. For example, while Crucell produces a range of vaccines and therapeutics, it may have a separate division for the pure licensing of its PER.C6® cell line. In fact, a major part of its business consists of the licensing of the cell line. In addition to the annual license payments, the licensees owe Crucell a royalty over the proceeds of any pharmaceuticals developed or manufactured with the cell line. Crucell can easily track whether the cell line has been so involved, as the drug registration authorities (FDA and EMEA) require all applicants to present a full description of all steps of the entire development and production process. Sample-shares could then, for instance, be designed so as to track to the financial performance of this particular licensing division. Again, there probably is not a single academic, public clinical or population biobank consent based governance system which grants its donors such extensive, flexible rights.

6.13 Complications

Both the issue of shares and the exercise of shareholders' rights could be complicated and raise new issues. For one thing, they create an administrative burden on the corporation, which has to register and keep track of its shareholders. This however, is not unique to holders of sample shares, and, as such, the hassle of their administration cannot provide an argument not to issue shares to such shareholders. Firms are capable of handling stock options plans for thousands of employees, so they might just as well be capable of handling a donor stock option plan, involving multiple donors. Furthermore, sample shares are likely to form a minority shareholding and as such will have to enjoy minority protections. Here too, however, the sample-shares being "ordinary" shares but for the fact that they have been issued to contributors of samples, will have the same (pre-emptive) rights (e.g. against dilution) as similar minority shareholders. Notably, in the press release announcing its bid for all outstanding shares in Crucell, JNJ stated that

---

[35]http://www.sec.gov/answers/track.htm (accessed 07 March 2011).

[36]Ibid.

> in accordance with customary Dutch practice, and to adequately protect the interest of any minority shareholders, Johnson & Johnson expects to retain two independent supervisory directors after closing.[37]

One problem may be that, in practice, sample shareholders may not be able to pay up on any newly to be issued shares in cash, which would make them vulnerable to dilution. This too, however, is a problem that is not unique for holders of sample shares and the mechanisms developed to protect against this type of risk could be deployed here as well. Another issue that could come up is the question of the transferability of the sample shares. If the sample shares are shares in a closed corporation with limited liability, (statutorily mandated) limitations to the transferability of these shares apply which aim to protect the closed character of the corporation. Depending on chosen form of the "blocking clause" in the articles of association of the corporation, these limitations will impose either a mandatory offering of the shares to the incumbent co-shareholders or will require the approval of the proposed transfer by a designated body of the corporation. Notably, these blocking clauses do not apply to transfer to a shareholder's spouse or relatives, an exception that sits nicely with the familial nature of the contributed sample: both samples and the corresponding sample shares can "stay in the family".

The question has been raised whether the model would also work in developing countries lacking a sophisticated legal and financial infrastructure. Probably not, but then most alternative proposals would not work in such countries either and it is hard to see why that would invalidate the model in countries with the proper infrastructure. Another criticism was that the model would have a built-in conflict of interests between the objective of the corporation to make a profit versus the objective of the sample-shareholders to see their samples used. While this may indeed be a conceptual problem, in practice, the objectives of the corporation and the sample donors are more likely to converge. Ultimately, the sample-shareholder will want his sample to be commercialised into products that he or others will actually benefit from (such as the Crucell vaccines) and thus products that are marketable and hence, by definition, profitable.

6.14 Discussion and Conclusion

Whether to ignore sample donor rights and focus on efficient use of the samples is better than identifying, articulating, and enabling the enforcement of these rights, is still open for debate. Crucell, for example, has provided great benefits to especially the developing world with a whole series of vaccines built on the PER.C6® cell line. Why would a donor of a mere sample want to share in the profits thereof? From an altruistic, economic, and even a legal point of view, a good case can be made that

[37] Johnson & Johnson and Crucell in advanced negotiations for an all cash public offer of €24.75 per ordinary share of Crucell, New Brunswick, NJ, and Leiden, The Netherlands, Press release 17 September 2010.

commercial use of human tissue, once excised, should be allowed to proceed without much ado.

On the other hand, human rights perspectives, utilitarian notions, and the unarticulated but profound sense that "you don't want to get screwed" by someone who has a free ride on your donation, call for some sort of benefit sharing. Fact is that many (potential) tissue donors see the (potential) commercialization of "their" tissue as highly problematic.[38] These concerns may act as a stumbling block towards commercialization, whereas such commercialization is widely recognized a necessity; not merely to allow biobanks holding those samples to become sustainable by exacting a commercial fee for access to their collections, but also for the more fundamental goal to transform the (technology and knowledge to be gained from) these samples into commercially viable and available applications.

While issuing shares for samples may seem an awkward fit, it may be a proper way to give sample donors a say in both the gains and the governance of their samples. Thus, a "shares for sharing" model would resolve a set of ethical-legal claims, in one stroke: the right to consent in the use of the tissue, the right to be informed about its further use and the right to share in any benefits resulting from its use. If there is consensus that research must be translated from bench to bedside and that this is to be done by the private sector, then the logical corollary is that the contribution of the raw material might as well be seen as a contribution in kind. In view of the widespread use in the biopharmaceutical industry of stock option plans (for employees, suppliers, accountants, lawyers), a share issue (or a donor stock option plan) to those who contribute indispensable material does not seem out of place. This is even more the case, where these contributors continue to provide new samples and/or (continuous) access to their corresponding health records.

An additional benefit of using a share based approach is that it bestows rights (both governance and gains) that are commensurate to the value of the contribution. Rather than providing tissue donors with a blanket power to unilaterally dictate the terms of commercialization without due regard to the contribution of other stakeholders, providing them with equity allows for fine-tuning and tailor made rights that reflect the proportionate value of this contribution in relation to the collaborative research enterprise as a whole. Further, the rights attached to an equity investment are time-tested and enforceable, and allow for a delicate balancing of power between the interests of all stakeholders in what has become the collaborative biomedical research enterprise. Notably, corporate law thus provides much more subtle and sophisticated instruments to resolve complex donor claims than the binary, one size fits all, informed consent doctrine has ever provided.

6.15 Epilogue

On 22 February 2011, Johnson & Johnson and Crucell N.V. announced that Johnson & Johnson had completed the tender offer for Crucell and had declared the offer unconditional. As Johnson & Johnson will hold at least 95% of the shares of Crucell

[38] Supra notes 8–12.

upon the settlement date, it intends to acquire the remaining shares not tendered by means of buy-out proceedings ("uitkoopprocedure") in accordance with article 2:92a and/or 359c of the Dutch Civil Code ...[39]

References

Australian Law Reform Commission (2004) "Essentially Yours: The Protection of Human Genetic Information in Australia." Accessed October 29, 2010. http://www.alrc.gov.au/events/speeches/DW/2004/20040226.pdf.

Bovenberg, J. A. (August, 2005) "Whose Tissue Is it Anyway." *Nature Biotechnology* 37: reprinted in: Bovenberg, J. A. (2006) reprinted in Brill/Martinus Nijhoff: *Property Rights in Blood, Genes and Data: Naturally Yours?* Leiden, and Boston: Brill/Martinus Nijhoff Academic Publishers. Leiden, Boston (www.brill.nl), 2006.

Gitter, D. M.. 2004. "Ownership of Human Tissue: A Proposal for Federal Recognition of Human Research Participants' Property Rights in Their Biological Material." *Washington and Lee Law Review* 61: 257.

Greely, H. T. 2000. "Iceland's Plan for Genomics Research: Facts and Implications." *Jurimetrics* 40: 153–91.

Kanellopoulou, N. 2011. "Reciprocity, Trust and Public Interest in Research Biobanking: In Search for a Balance" In *Human Tissue Research - A Discussion of the Ethical and Legal Challenges from a European Perspective*. edited by C. Lenk, N. Hoppe, K. Beier, and C. Wiesemann, 197–218. Oxford: Oxford University Press.

Laurie, G. 2002. *Genetic Privacy: A Challenge to Medico-Legal Norms*. Cambridge: Cambridge University Press.

Potts, J. 2002. "An Examination of the Bargain Made Between Iceland and deCODE Genetics with Implications for Global Bioprospecting." *Virginia Journal of Law & Technology* 7: 8.

Skloot, R. 2010. *The Immortal Life of Henrietta Lacks*. New York: Crown Publishers.

Terry, P. F. 2003. "PXE International: Harnessing Intellectual Property Law for Benefit-Sharing." In *Populations and Genetics*, edited by B. M. Knoppers, 377–93. Leiden and Boston: Martinus Nijhoff Publishers.

Winickoff, D. E., and R. N. Winickoff. 2003. "The Charitable Trust as a Model for Genomic Biobanks." *New England Journal of Medicine* 12: 349.

[39]Press release, Johnson & Johnson Completes Tender Offer for Crucell and Declares Offer Unconditional, New Brunswick, NJ, and Leiden, the Netherlands, 22 February 2011.

Chapter 7
Research on Human Biological Materials: What Consent Is Needed, and When

Eugenijus Gefenas, Vilius Dranseika, Asta Cekanauskaite, and Jurate Serepkaite

7.1 Introduction

Requirement for informed, express and specific consent is one of the key principles of research ethics that evolved as a reaction to the atrocities of the Nazi medicine as well as a response to the unethical human experimentation revealed during the post World War II period. Such consent is thought to be the default position in clinical research and any softening of the requirement is usually perceived as an exception which requires justification. However, in some areas of human research the requirement is more and more often both weakened in practice and criticised by the members of research community and ethicists (e.g. Chadwick and Bere 2001; Hansson et al. 2006; Helgesson et al. 2007). In this chapter we discuss circumstances under which research on human biological materials is in fact conducted without specific consent or re-consent of a donor.

We start our discussion with a short note on the relationship between consent and identifiability of human biological materials. The main part of the chapter is concerned with issues of consent in research on identifiable archived human biological materials. This covers three main scenarios. First, we explore the research use of materials that were collected for broadly-defined research purposes for which broad consent was initially secured. Second, we discuss the possibility to waive consent in research use of biological materials that were initially collected for non-research purposes without consent for research use. Third, we address three alternative regulatory regimes allowing turning residual biological materials into research collections during the collection procedure. These alternatives that justify research use of biological materials collected for non-research purposes can be based on precautionary consent, presumed consent and no consent. Before drawing conclusions we will also briefly discuss the removal of biological materials for research purposes from the deceased.

E. Gefenas (✉)
Department of Medical History and Ethics, Vilnius University, Vilnius, Lithuania
e-mail: eugenijus.gefenas@mf.vu.lt

C. Lenk et al. (eds.), *Biobanks and Tissue Research*, The International Library of Ethics, Law and Technology 8, DOI 10.1007/978-94-007-1673-5_7,

Our scope in this chapter is mainly European. However, there is no unified legally binding European policy in relation to any of these fields of research and different countries often set different regulations. Therefore we face the risk of presenting the situation as more uniform and orderly than it really is. While recognising these risks, we hope to describe the general landscape and identify some of the emerging trends by referring to some examples taken from a number of European countries.

7.2 Identifiability and Non-Identifiability

The possibility to link human biological material to a particular person from which this material was taken is an important feature that influences the stringency of expected consent. The more difficult it is to establish this link, the more it is likely that softer consent requirements will be applied. This general observation seems to be very simple but the real-life situations are not as straightforward as this. A number of authors have observed the lack of common terminology in regulatory documents concerning identifiability of human biological materials (e.g., Knoppers 2005; Knoppers and Saginur 2005; Elger and Caplan 2006; van Veen et al. 2006). The reader should be cautious when such terms as "anonymous", "coded", "anonymised", "unlinked", "double-coded" etc. are encountered in the literature. The same biological materials may be called anonymous in one jurisdiction and identifiable in another. For example, Elger and Caplan (2006) have shown some systematic differences between Europe and the United States of America (USA) both in interpretation and application of such notions. If coded samples are provided to the researcher in a form that does not allow the identification of the donor, they are treated as anonymous samples in the USA, but as identifiable in many of the European jurisdictions. Therefore, it is important to be clear from the outset concerning the terminology used.

We will use the term "non-identifiable" in a sense that is given in the Council of Europe Recommendation Rec(2006)4 of the Committee of Ministers to Member States on research on biological materials of human origin, that is "those biological materials which, alone or in combination with associated data, do not allow, with reasonable efforts, the identification of the persons concerned." Identifiable materials, on the other hand, are defined as those "which, alone or in combination with associated data, allow the identification of the persons concerned either directly or through the use of a code" (Council of Europe 2006a: Art 3).

It is important to keep in mind that even in Europe the term "anonymous" is sometimes used to refer to materials that are identifiable in a sense provided by the Rec(2006)4. For example, materials which are stored in a repository in an identifiable format but are surrendered to the researcher in a form that does not allow him or her to identify the donor are sometimes referred to as "anonymous" whereas the Rec(2006)4 employs the term "linked anonymised materials" in such circumstances. The term "anonymous to the researcher" can also sometimes be encountered. Most jurisdictions within the European Union would not treat such data as anonymous, but some would (e.g. the UK, the Netherlands, Austria) (van Veen et al. 2006: 2917).

It also seems important to note the fact that even though the Rec(2006)4 introduces a distinction between "coded" (researcher has an access to the code) and "linked anonymised" (the code is under control of a third party) materials, it does not define separate regimes for the treatment of the samples of these two kinds. However, this distinction seems to be more pronounced in the national law of a number of countries.

Research on human biological materials that have already been collected and are treated as anonymous in a given jurisdiction can in many cases proceed without consent, as, for example, indicated in Norwegian Health Research Act ("Consent is not required for research on anonymous human biological material and anonymous data" (Parliament of Norway 2008: Ch 4, Sec 20)) or supported by such national bodies as the German National Ethics Council ("If samples and data lawfully obtained for diagnostic or therapeutic reasons are subsequently used for medical research, the requirement of consent may be waived if the samples and data are completely anonymised.") (2004: 12) and the French National Consultative Ethics Committee for Health and Life Sciences ("It would conceivably be acceptable that should personal data be scrupulously anonymised, it could be used for subsequent research without renewing consent procedures.") (French National Consultative Ethics Committee for Health and Life Sciences 2003: IV.2). Terminological difficulties can once again be clearly seen in different wordings employed in different documents. Is such a notion as "impossibility to identify with reasonable efforts" less restrictive than "complete anonymisation" or "scrupulous anonymisation"? Consequently, a careful legal analysis may be needed in order to establish whether in these cases it is non-identifiable materials in a sense provided by the Rec(2006)4 are being addressed or a broader category that in addition to the non-identifiable materials would also cover some types of identifiable materials.

Even though there is a general tendency to allow research on non-identifiable materials without consent, the situation becomes a bit more complicated when the person removing the material is also a researcher who is going to conduct research on that material, because in these circumstances the donor can almost always be identified. Therefore, a question can be raised in this context about the moment when the materials removed from the human body become non-identifiable? Does this happen when the materials are not labelled with the identifiers of any sort during the removal procedure or should some additional measures be also taken?

Finally, research use of non-identifiable materials may not require consent, but this does not apply to the removal of non-identifiable materials. Where materials are taken for specific research projects or broadly defined future research purposes (and, say, immediately rendered non-identifiable) there is always a contact with the donor and consent is required.

7.3 Prospective Research Collections and Broad Consent

Where prospective research is concerned, the practice of collecting identifiable human biological specimens with broad consent for research use is becoming more and more common (e.g. Wendler 2006; Hansson et al. 2006; Petrini 2010). It is also recognised as acceptable by a number of international organisations, such as the Council of Europe (2006b: Sec 42) and the Organisation for Economic Co-operation and Development (2009: Part II, Sec 28). By "broad consent" we mean any consent that is broader than specific consent to use samples in a particular research project. Broad consent may come in degrees ranging from consent to conduct research in a specific field of biomedicine or on specific disease, such as pharmacogenetic studies or most of the disease oriented biobanks, to unrestricted consent for any future research use, as is usually the case in large-scale population biobanks. Let us illustrate this by providing a list of formulations of consent differing in breadth (in ascending order):

- one particular study with no further contact permitted (specific consent);
- one particular study with permission for further contacts to conduct further studies;
- studies of one particular type with no further contact permitted;
- studies of one particular type with permission for further contacts to do further studies of other types;
- any kind of future studies (unrestricted consent).

It is important to note that in some cases, more than one of the listed options may be offered to the donors. This would constitute the so-called multi-layered consent (Salvaterra et al. 2008).

Where broad consent for the use of prospectively-collected specimens is available, research ethics committees (RECs) still need to evaluate whether a specific research protocol is in line with the initial consent. This may be unproblematic in biobanks where very broad, almost unrestricted consent is secured. For example, the Estonian GeneBank gene donor consent form simply states that "The Gene Bank enables scientific and applied gene and health research to be carried out in order to determine genes that influence the development of diseases." (Estonian Ministry of Social Affairs 2007). However, current regulations tend to discourage unrestricted consent (e.g. so-called 'blanket consent' in Draft Explanatory Memorandum to Rec(2006)4 (Council of Europe 2006b: Sec 48) or "overall prior consent" referred to in the Report on Consent by UNESCO (2008: Sec 157)). For example, the UNESCO report states that "It is not acceptable to ask a participant in a research project to give overall prior consent (so-called 'blank consent') to the effect that they would agree to any study that can be carried out with the data/material they provided, unless the data/material be irretrievably unlinked to the participants" (ibid.). Such a consent may be almost unavoidable in large scale population biobanks (e.g. Estonian GeneBank or UK Biobank) but in cases of smaller and more focused biobanks it is often possible to rely on consent that is broader than specific but narrower than the unrestricted one. However, even though research collections relying on broad consent may be unable to provide the donors with the information on potential future research uses they may be able to inform them on a number of procedural issues such as models of governance, existing data protection measures, and donor rights (e.g. Estonian Ministry of Social Affairs 2007; UK Biobank 2010).

In case of prospective research collections there seems to be an emerging trend to seek a compromise between specific consent and unrestricted overall consent. Even though it is often deemed acceptable to secure broad consent while collecting materials for prospective collections, the consent should not be broader than necessary (e.g. Porteri and Borry 2008: 139). As stated by Carlo Petrini, "'broad consent' is not like signing a blank check" (2010: 220). The Norwegian legislation can serve as an example of this tendency: the recent Norwegian Health Research Act states that "research participants may consent to human biological material and personal health data being used for specific, broadly defined research purposes" (Parliament of Norway 2008: Ch 4, Sec 14).

When judging whether the new use is in line with the initial consent, it is important to keep in mind the variety of levels of consent in terms of breadth. Thus the decision matrix is the following: is the new research proposal in line with the initial consent? If yes, there is no need to re-contact. If no, the situation is in all important respects analogous to the one that will be described in the next section on secondary use of archived materials where no initial consent was secured.

7.4 Waiving Consent in Secondary Research Use of Human Biological Materials

Collections of human biological materials taken for non-research purposes such as diagnostics, public health screening, or quality assurance, may be found in a variety of healthcare institutions including hospitals, pathology laboratories, tissue banks, blood banks, genetic laboratories etc. This is still "one of the richest sources of tissue for research" (Bathe and McGuire 2009). Such repositories may store human biological materials in different forms and different regulations may apply during different stages of processing and storage of the materials. For example, in case of a pathology laboratory, biological materials are taken for diagnostic purposes and later they are processed and stored in different forms that can be used for laboratory diagnostics. Such forms may include formalin-fixed paraffin-embedded tissue blocks (FFPE), tissue microarrays, or nitrogen-frozen tissue, which is particularly suitable for the extraction of RNA. It is also important to note that after the diagnostic sample is prepared, the rest of the biological material becomes a leftover material that is most often processed as a medical waste. Usually there is a set period during which the institution is required to keep the diagnostic samples for possible subsequent use. When this period expires, the samples may either be destroyed or stored further. In some cases these materials may be interesting for scientific research. However, research use is often not covered in the consent procedures during the obtaining of these samples. Figure 7.1 provides an example of different forms of diagnostic samples and different stages at which biological materials can be used for research purposes in a typical pathology laboratory in Lithuania. The storage times, requirements for ethical review, sample preservation technologies etc. may differ across countries, however, the structural elements of the process remain almost the same in different countries.

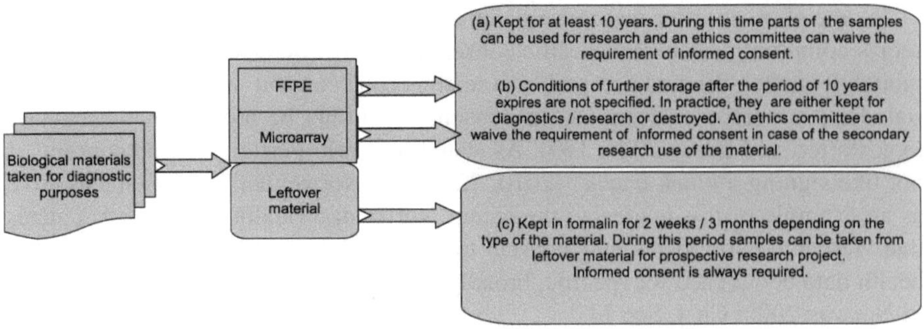

Fig. 7.1 An example of the flow of samples in a pathology laboratory in Lithuania

It is quite common to think that re-contacting of donors is the preferable ethical policy in case of secondary use of biological materials. However, it is often recognised that in case of research on large-scale collections of identifiable archived human biological materials, re-contacting might not always be practicable and the waiving of consent is considered as an acceptable policy (e.g. Knoppers and Saginur 2008; Tarini et al. 2008). Regulations may designate bodies, usually RECs, which are granted a power to waive the consent requirement under certain conditions.

This is explicitly stated in some international and European regulations. For example, the Declaration of Helsinki states that "There may be situations where consent would be impossible or impractical to obtain [...]. In such situations the research may be done only after consideration and approval of a research ethics committee" (World Medical Association 2008: Art 25). Another important document – Rec(2006)4 – states that biological materials can be used in a research project without the consent of the person concerned when an independent evaluation certifies that contacting the person concerned is not possible with reasonable efforts, and (a) the research addresses an important scientific interest; (b) the aims of the research could not reasonably be achieved using biological materials for which consent can be obtained; (c) there is no evidence that the person concerned has expressly opposed such research use (Council of Europe 2006a: Art 22).

Similar provisions can be found in national regulations as well (e.g. Finnish Act on the Medical Use of Human Organs and Tissues (Parliament of Finland 2001: Sec 20); Norwegian Health Research Act (Parliament of Norway 2008: Ch 6, Sec 28)). Sometimes the right to waive the consent requirement is granted to RECs but no further guidance is provided on how the RECs may base their decision. For example, the Lithuanian Law of Ethics of Biomedical Research (Seimas of the Republic of Lithuania 2000: Art 8.2) simply states: "Whether the subject's informed consent is necessary [...] shall be decided by the Lithuanian Bioethics Committee or the Regional Biomedical Research Ethics Committee giving their approval."

Situations in many respects analogous to the one previously described, can be encountered when human biological materials collected with specific consent for particular research projects are later being utilised for other research purposes. Quite obviously, initial consent, provided that it was specific and no future research use or research storage was mentioned, does not imply consent for secondary use.

Where RECs are given the power to waive the consent requirement they should evaluate whether, under given circumstances, conducting a particular research is more important than securing consent. A number of reasons to waive the consent requirement can be encountered in the literature of which the most often cited one is difficulty in contacting the donors. There are two possible approaches to the problem. Conditions under which the waiver is deemed to be possible may simply include a relative notion, such as "unreasonable", "impractical", "non-practicable" (e.g. World Medical Association 2008; UNESCO 2008). In addition to such a relative notion they may also include a list (sometimes open) of considerations which may be deemed to justify the waiver (e.g. Council of Europe 2006a; Finnish Act on the Medical Use of Human Organs and Tissues (Parliament of Finland 2001: Sec 20)). In both cases, institutions that in fact have to make decisions on particular cases are not given a decision procedure or necessary guidance as to what may be considered a good decision.

Even though no exhaustive guidance exists in this field, let us list a few of the different reasons which may be considered as rendering contacting the donors "unreasonable" (adapted from Junod and Elger (2010: E4)):

- It may be too costly because the donors may have moved to distant areas, there are no easy methods of identification of their whereabouts, and the number of samples is very big.
- It may be impossible since there is reason to believe that the donors may not be alive anymore.
- It may cause distress in patients (especially in cases of psychiatric illnesses) or their relatives (e.g. where patient may have died in dramatic circumstances).
- It may have a risk of stigmatisation if there are chances that other people will receive the information.

In addition to these, there may be other important considerations. Adequacy of coding and data protection procedures may serve as an argument to waive the requirement to secure consent. It is also important to stress that the scientific merit of the research in question and possible utilisation of research results is another crucial consideration. Scientific merit is both *conditio sine qua non* for ethical approval of research and it may serve as an additional argument to outweigh the necessity to secure the re-consent of the research subjects.

Enumeration and discussion of the mentioned conditions is important since this may help to clarify the notion of "reasonability". However, this does not amount to clear guidelines for RECs. For example, if the number of samples is concerned, what amount of samples is large enough to justify the waiver? And if the number is set, does it mean that in cases of smaller numbers of samples, consent is always needed? Should researchers' financial abilities be taken into account when deciding whether the burden will be impracticable? The problem of lack of guidance has been explicitly noted by David Hunter (unpublished manuscript) who observes that the new UK Tissue Act established a system where RECs are given an obligation to assess whether the consent requirement can be waived, but are given almost no guidance on how this should be done. It is important to ask whether placing an obligation on RECs to judge if consent is needed and not providing them with guidelines is the best policy. It is very important to learn more about actual decisions and decision-making processes in order to evaluate this policy.

It is difficult to tell how often RECs decide to waive the consent requirement. It is perfectly reasonable to ask whether, though consent is treated as a default and preferable position in legal documents and among ethicists, in practice, it is the waiver of consent that is the default position and refusals to waive the consent are relatively rare. As noted by Metti Hartlev and Uffe Lind, this may be the case in Denmark: "In the research ethics committee system's practise, derogation [from consent] seems to be the rule rather than the exception." (2006: 8). It would be important to investigate this issue more thoroughly as it may suggest certain discrepancies between ethical regulations and actual practice.

7.5 Alternative Strategies for Secondary Use of Biological Materials: Turning Residual Materials Into Research Collections

In the previous section we discussed the scenario of secondary research use of biological materials where waiving of consent is one possible scenario to overcome regulatory "obstacles" to such a research. Some countries, however, have chosen another strategy to bypass the complexities of regulatory requirements as applied to research activities on residual biological materials. This strategy is simply based on the manoeuvre to turn biological material into the research collection right during the collection procedure. It can be employed when, to use the terminology of the Rec(2006)4, residual biological materials are collected during interventions carried out "for purposes other than storage for research" (Council of Europe 2006a: Art 12.1). The goal to use the residual materials obtained during therapeutic or diagnostic interventions for future research can be achieved by three different options: (a) broad, precautionary consent may be secured before collecting the materials; (b) the presumed consent model may be applied during the collection of materials; and (c) consent for research use of identifiable human biological materials may be skipped at all. These three options differ in their legal handling of consent issues but are rather similar in their practical implications.

(a) The first option overlaps with the scenario described in the section on the broad consent in prospective research collections as it is based on consent secured before or during the process of the routine collecting of the residual materials. For example, in its opinion of 2004 the German National Ethics Council recommends to seek "precautionary consent" for future medical research purposes even where the bodily substances are collected for diagnostic or therapeutic purposes. This allows avoiding the issue of re-consent if at some later point the need to use the samples for research purposes arises (2004: 12). Since future research uses cannot be known with any precision at the moment of "precautionary" consent, we suppose that such a consent would in this situation be framed in rather broad terms. Precautionary consent is also envisaged by the Council of Europe, albeit not using the term "precautionary consent". In relation to the practice of collecting residual materials the Explanatory Memorandum to the Rec(2006)4 recommends that "When biological materials of human origin and personal data are collected it is best practice to ask the sources for their consent to future use, even in cases where the specifics of the future research projects are unknown" (Council of Europe 2006b: Sec 48). A similar position is expressed by the Irish Council for Bioethics recommending to routinely ask the patients for consent for the possible future research use of "tissue or organs removed during surgical treatment or surplus biological material left over after diagnostic testing." The Council took into consideration that "[s]uch consent by its nature would have to be a broad consent, as the type of the research is totally unknown at the time [when it's taken from the subject]." (2005: 31–2).

It is important to note that in the context of collecting residual human biological materials for future research precautionary consent does not have to be construed as a totally unrestricted consent. In some cases it may be possible to specify potential future research uses to some extent (e.g. using the multi-layered consent described in Sec. 3) and then to ask the donor to choose the breadth of consent.

(b) In some countries, identifiable biological materials removed during diagnostic or therapeutic procedures may be retained without express consent for research use, and later these collections may be treated as already-existing research collections. In order to make this possible, some countries have introduced general opt-out schemes for use of residual tissues anonymous to the researcher.

Such a system may be said to operate in the Netherlands. This country introduced a general opt-out scheme for the use of residual tissues anonymous to the researcher. The Dutch Code for Proper Secondary Use of Human Tissue describes mechanisms of presumed consent: "For the use of coded or anonymous material [for scientific research] it is sufficient if the subject (donor) has not objected to this use." (Federation of Medical Scientific Societies 2003: Art. 4.2). It is important to note that "coded" here means "linked-anonymised" in a sense provided by the Rec(2006)4. The Dutch Code explains that "The choice system assumes that considerable informative material will be made available in the institution where the human material is collected for original use. The informative material states that 'further use' of human material does occur, in particular for scientific research" (Federation of Medical Scientific Societies 2003: Art 4). A similar system has been recently introduced in Belgium (Belgium Federal Parliament 2008: Art 20.2), Denmark (Danish National Committee on Biomedical Research Ethics 2009: 2.6.2) and it is also recommended by the Austrian Bioethics Commission at the Federal Chancellery (2007).

The Dutch, Belgian, Austrian, and Danish examples are interesting since the rationale behind them seems to be the extension of the opt-out regimes that are usually used for transplantation policies to the field of biomedical research. An opt-out system is based on the assumption that the population in general is not opposed to the research use of such materials, and that it is sufficiently informed of the fact that such use may occur. It should be noted, however, that in Austria and Belgium the suggested opt-out regime applies only to the left-over residual materials and does not extend to the biological samples archived for diagnostic purposes (Belgium Federal Parliament 2008; Bioethics Commission at the Federal Chancellery 2007).

(c) The most straightforward strategy would be not to require consent at all but we are not aware of any European examples that would fit it uncontroversially. Perhaps the closest example is the UK Human Tissue Act (2004) which does not require consent for storage and research use of residual tissues that are either fully anonymous or anonymous to the researcher. However, such research can proceed only if there is a permission from the REC (Parliament of the United Kingdom 2004: Art 1, Sec 7–9). Here permission cannot be interpreted as dealing with the waiver of consent since consent is not needed at all. Even though the UK Tissue Act requires no consent, the Code of Practice issued by the Human Tissue Authority (HTA) while also allowing research with no consent, calls the situation where the consent is secured "the preferable scenario" (Human Tissue Authority 2009: Sec 26). The Code of Practice also provides a different option to start research activities without consent, namely, to get one-time ethical approval by the recognized REC for research use of tissue collection stored in a particular HTA-licensed tissue bank. Provided that such an approval is secured, individual research projects do not need further REC approvals but the tissue bank is required to work under standard operating procedures (SOPs) issued by the National Research Ethics Service (Human Tissue Authority 2009: Sec 68).

These cases present a significant interest here because establishing a system that does not require consent, presumes it, or allows to routinely obtain a precautionary one, secures a constant and ample supply of identifiable human biological materials for research. However, bearing in mind that explicit consent for collection of human biological materials for future research purposes is still required by most of the international documents, the availability of the alternative regulatory regimes that allow turning residual biological materials into research collections during the collection procedure marks a significant change in attitude towards the requirement of consent. Although this trend is suggested by some authors (Coebergh and his colleagues (2006) and van Veen (2006)), it should be stressed that the mentioned alternative options could also rise serious ethical concerns related to the protection of the interests of the donors.

7.6 Removal of Biological Materials from the Deceased for the Research Purposes

Let us briefly address the issue of research use of materials taken from the deceased. The national regulatory frameworks here are rather diverse. Legal regulations range from the restrictive positions that disallow the use of anonymous materials without expressed consent of the deceased or his or her relatives to the possibility to remove even the identifiable post-mortem materials without explicit consent.

The UK can be seen as an example of the country that has one of the strictest consent regulations on the research use of post-mortem biological materials. It maintains that the removal or use of biological materials from the deceased should always be conditional upon appropriate consent (either expressed by the person before death or given by a nominated representative, or in absence of either of these, given by a legal representative). It is interesting to note that this strict regime has replaced a more liberal regime following a major scandal involving the unauthorised removal, retention, and disposal of human tissue (Price 2005). As has been mentioned earlier, in case of the living donors, the UK does not always require consent, i.e. if the sample is anonymous to the researcher and respective ethics committee has approved this research project (Parliament of the United Kingdom 2004: Sec 1). It seems, therefore, that at least in some cases (see (c) of Section 7.5) the stricter consent regulations are applied to the research use of post-mortem materials than to the research use of materials taken from the living in the UK.

Other European countries tend to soften consent regulations in relation to the biological materials removed from the deceased. For example, it is permitted to use anonymous samples for research purposes without consent in Estonia (Parliament of Estonia 2005). The Finnish Act regulating Medical Use of Human Organs and Tissues contains the provision that permits taking both anonymous and identifiable samples without explicit consent if there is an approval of a special institution titled the National Board of Medicolegal Affairs (Parliament of Finland 2001: Sec 11, 12).

Spain (Uranga et al. 2005) and France (Knoppers and Saginur 2008) seem to follow the presumed consent model. The model of presumed consent with some limitations is probably going to be also established in Switzerland. The draft of Swiss federal law on human research includes consent provision stating that when the biological materials are removed during the autopsy or transplantation, a small quantity of biological materials can be anonymised for research if there is no dissent document (Swiss Parliament 2009: Art 37).

The Latvian Law on the Protection of the Body of a Deceased Human and Utilisation of Human Tissues and Organs in Medicine might be considered as one of the least restrictive examples of consent regulations. It provides for a possibility to obtain the samples for research purposes during pathological-anatomical and forensic examination "if the deceased has permitted it during his or her life, if the will of the deceased is unknown, as well as if the will of the deceased is unknown and he or she does not have the next of kin" (Parliament of Latvia 1992: Sec 9). It seems, therefore, that the removal of tissue and cell samples is forbidden only if there is evidence that the deceased has objected to this during his/her life.

Some countries do not have specific consent regulations regarding the research concerned. For example, the Lithuanian Law on Ethics of Biomedical Research (which includes research on human biological materials into its scope) contains a general consent requirement for any kind of biomedical research, which leads to the situation where the only possibility to use the samples of the deceased is his or her will (consent) expressed while being alive (Seimas of the Republic of Lithuania 2000: 8.1).

Each of these strategies are compatible with the Rec(2006)4, which does not specify consent requirements for removing samples from the deceased for research because it only states that samples cannot be removed without either consent or authorisation, and that they cannot be removed and used for research activities if a person is known to have objected to it (Council of Europe 2006a: Art. 13.1-2).

We would like to conclude by noting that there seems to be no unified European consent policy and no clear emerging trends in relation to this field of research. Different countries set different consent models. Some countries always require relatives' consent, the others implement presumed consent model or set conditions under which consent can be waived, presumed, or not required at all. It is also worthwhile to discuss to what extent such regulatory disparities have a potential to hinder international research in Europe.

7.7 Conclusions

Research on archived identifiable human biological materials is an important area of biomedical research. However, strict requirements on consent are sometimes seen as an obstacle to this type of research. A number of conditions have been specified in international and national regulations which permit the softening of the consent regime in different types of research collections. Two most often used scenarios have been presented in this chapter. First, the strategy to broaden consent in prospective research collections, especially large scale, population based biobanks. Second, secondary research use of human biological material combined with the power given to RECs to decide whether the requirement can be waived in a particular case. This scenario is very important due to its potential to provide material for disease based collections. The third scenario presented in the chapter could be regarded as one of the alternatives to justify the secondary use of biological materials. This scenario, which seems to be the least discussed option, is based on the strategy to turn residual materials into research collections right at the outset of the collecting procedure The third scenario covers some of the most permissive regimes such as presumed consent and no consent. These strategies may present the easiest way to make large amounts of human biological materials available for research. Finally, issues of consent in removal of biological materials for research purposes from the deceased are regulated in very different ways across Europe.

Various possible shifts from the specific consent regime are summarized in Table 7.1:

Table 7.1 Shifts from the specific consent regime in research on different types of collections of human biological materials

Type of collection	Shift from the specific consent
Anonymous collections	In many cases no consent is needed. However, which collections are called "anonymous" differs across jurisdictions.
Prospective research collections	A tendency towards broadening of the consent – compromise between specific consent and unrestricted, overall consent.
Secondary use of non-research collections of residual materials	Waiver of consent requirement is acceptable. However, delineation of circumstances under which it is possible is problematic.
Turning residual materials into research collections during the collection procedure	A number of strategies in different countries: (a) broad precautionary consent; (b) presumed consent; (c) no consent needed.
Materials removed from the deceased	No clear trends.

While discussing different strategies of handling consent issues in human tissue research it is important to compare regulatory stringency in different types of research. At least in some cases regulatory frameworks dealing with prospective collections are very comprehensive and detailed. These regulations seek to balance or compensate the broadening of informed consent with specific legislation and sophisticated project management and personal data protection systems which in case of large-scale population biobanks may even include specially assigned ethics bodies. Secondary research use of biological materials in many cases is regulated less strictly and systematically. Especially because waiving of informed consent is usually only balanced by the requirement of ethical review of a particular project and taking into account that waiving of consent increasingly becomes the default policy among RECs. Finally, the third scenario includes some of the most diverse ways to balance the softened model of consent. This scenario may be even based on policies prescribed by the guidelines of professional organisations rather than legal regulations, as it is the case in the Netherlands. In addition, most of the organizational structures related to this scenario are limited to a particular health care institution conducting research. Remarkable differences in regulatory stringency associated with different types of human tissue research are hardly justifiable having in mind that all these types fall within the broad field of non-interventional and minimal-risk research. For example, Klaus Hoeyer points out to this type of regulatory discrepancy in Denmark where the opt-out system for routine tissue storage "has created a strange system of double standards: no consent is needed for using a tissue sample for research if it is taken for diagnostic purposes and used for research only at a later stage; while samples taken specifically for research must be collected with consent" (Hoeyer 2008).

Softening of the consent procedures provides scientists with easier access to human biological materials, but also raises a number of important ethical concerns. These concerns include those of privacy, especially relating to the security and efficiency of coding procedures, and the possible failure to protect autonomy of the donors. We would like to conclude the chapter by mentioning some additional themes that could supplement the analysis of different consent regimes in human tissue research. First, recognising that theoretical analysis of these questions is very important it may also be useful and important to take into account attitudes and preferences of various interested parties. A number of empirical studies were conducted during which patients were asked about their preferences concerning consent procedures in donating tissues for human tissue research (e.g. Stegmayr and Asplund 2002; Furness and Nicholson 2004; Wendler 2006; Vermeulen et al. 2009). The studies tend to show that in general patients highly endorse tissue research and agree that their samples be used for the future studies. However, it is also important to study which consent mechanisms are perceived to save their interests best. For example, Vermeulen et al. (2009: 1172) report a study in the Netherlands in which 56% of the respondents favoured a "one-time general consent" and only 23% preferred the current "opt-out" procedure. It is also important to investigate some other issues, such as opinions on commercial research or patients' wishes to be informed on the uses of their biological samples and research findings. Second, after delineating some trends in a descriptive manner we think it is also important to stress that some types of research may need stricter consent regulations than the others and thus softening of the specific consent regime, if needed at all, does not imply uniformity of regulations across all types of research. Such issues as research on materials obtained from children or other vulnerable populations may pose additional risks and thereby can justify additional protections.

References

Bathe, O. F., and A. L. McGuire. 2009. "The Ethical Use of Existing Samples for Genome Research." *Genetics in Medicine* 11: 712–15.

Belgium Federal Parliament. 2008. "Kingdom of Belgium: Law on Procurement and Use of Human Bodily Material for Medical Application to Humans and for Scientific Research." Accessed 08 March 2011. Available in French at: http://www.ejustice.just.fgov.be/cgi/article_body.pl?language=fr&caller=summary&pub_date=2008-12-30&numac=2008018385#top

Bioethics Commission at the Federal Chancellery. 2007. "Biobanks for Medical Research." Accessed 08 March 2011. http://www.bka.gv.at/DocView.axd?CobId=25510

Chadwick, R., and K. Bere. 2001. "Solidarity and Equity: New Ethical Frameworks for Genetic Databases." *Nature Reviews Genetics* 2 (4): 318–21.

Coebergh, J. W. W., E. B. van Veen, J. P. Vandenbroucke et al. 2006. "One-Time General Consent for Research on Biological Samples – Opt Out System for Patients is Optimal and Endorsed in Many Countries." *British Medical Journal* 332 (7542): 665.

Council of Europe. 2006a. "Recommendation Rec(2006)4 of the Committee of Ministers to Member States on Research on Biological Materials of Human Origin." Accessed 08 March 2011. https://wcd.coe.int/ViewDoc.jsp?id=977859

Council of Europe. 2006b. "Draft Explanatory Memorandum to the Draft Recommendation Rec(2006) of the Committee of Ministers to Member States on Research on Biological

Materials of Human Origin." Accessed 08 March 2011. https://wcd.coe.int/ViewDoc.jsp?id=961161&Site=COE

Danish National Committee on Biomedical Research Ethics. 2009. "Guidelines About Notification etc. of a Biomedical Research Project to the Committee System on Biomedical Research Ethics." Accessed 08 March 2011. http://www.cvk.sum.dk/English/guidelinesaboutnotification.aspx.

Elger, B. S., and A. L. Caplan. 2006. "Consent and Anonymization in Research Involving Biobanks: Differing Terms and Norms Present Serious Barriers to an International Framework." *EMBO Report* 7: 661–66.

Estonian Ministry of Social Affairs. 2007. "Republic of Estonia. Annex 1 to the Decree No 36 of the Ministry of Social Affairs of Estonia: Gene Donor Consent Form." Accessed 08 March 2011. http://www.geenivaramu.ee/index.php?id=100

Federation of Medical Scientific Societies. 2003. "Code for Proper Secondary Use of Human Tissue in the Netherlands." Accessed 08 March 2011. http://www.federa.org/?s=1&m=82

French National Consultative Ethics Committee for Health and Life Sciences. 2003. "Opinion n° 77 – Ethical Issues Raised by Collections of Biological Material and Associated Information Data: 'Biobanks', 'Biolibraries'." Accessed 08 March 2011. http://ec.europa.eu/research/biosociety/pdf/opinion_77.pdf

Furness, P. N., and M. L. Nicholson. 2004. "Obtaining Explicit Consent for the Use of Archival Tissue Samples: Practical Issues." *Journal of Medical Ethics* 30: 561–64.

German National Ethics Council. 2004. "Opinion on Biobanks for Research." Accessed 08 March 2011. http://www.ethikrat.org/_english/publications/Opinion_Biobanks-for-research.pdf

Hansson, M., J. Dillner, C. Bartram et al. 2006. "Should Donors Be Allowed to Give Broad Consent to Future Biobank Research?" *The Lancet Oncology* 7: 266–69.

Hartlev, M., and U. Lind. 2006. "Use of Blood Samples from the Danish PKU-Biobank – A Study of the Conceptualization of Research in Law and Action." Accessed 08 March 2011. http://www.eureca.manchester.ac.uk/newsletters/cases/thefile,116174,en.pdf

Helgesson, G., J. Dillner, J. Carlson et al. 2007. "Ethical Framework for Previously Collected Biobank Samples." *Nature Biotechnology* 25: 973–76.

Hoeyer, K. 2008. "An Opt Out System for Tissue Storage: Lessons from Denmark." *British Medical Journal*. Accessed 08 March 2011. http://www.bmj.com/content/337/bmj.a345.abstract/reply

Human Tissue Authority. 2009. "Code of Practice 9: Research." Accessed 08 March 2011. http://www.hta.gov.uk/legislationpoliciesandcodesofpractice/codesofpractice/code9research.cfm?FaArea1=customwidgets.content_view_1&cit_id=764&cit_parent_cit_id=757

Hunter, D. (Unpublished manuscript) *Tis but a Scratch: The Human Tissue Act and the Use of Tissue for Research, Issues for Research Ethics Committees*.

Irish Council for Bioethics. 2005. "Human Biological Material: Recommendations for Collection, Use and Storage in Research." Accessed 08 March 2011: http://www.bioethics.ie/uploads/docs/BiologicalMaterial.pdf.

Junod, V., and B. Elger. 2010. "Retrospective Research: What Are the Ethical and Legal Requirements?" *Swiss Medical Weekly* 140: w13041.

Knoppers, B. M. 2005. "Biobanking: International Norms." *Journal of Law Medicine and Ethics* 33 (1): 7–14.

Knoppers, B. M., and M. Saginur. 2005. "The Babel of Genetic Data Terminology." *Nature Biotechnology* 23: 925–27.

Knoppers, B. M., and M. Saginur. 2008. "Bio-Banking." In *The Cambridge Textbook of Bioethics*, edited by P. A. Singer and A. M. Viens, 166–73. Cambridge: Cambridge University Press.

Organisation for Economic Co-operation and Development. 2009. "OECD Guidelines on Human Biobanks and Genetic Research Databases." Accessed 08 March 2011. http://www.oecd.org/dataoecd/41/47/44054609.pdf

Parliament of Estonia. 2005. "Republic of Estonia. Law No. 821 Regarding the Determination of Death's Cause." Accessed 08 March 2011. Available in Estonian: https://www.riigiteataja.ee/ert/act.jsp?id=895466

Parliament of Finland. 2001. "Republic of Finland. Act No 101/2001 on the Medical Use of Human Organs and Tissues." Accessed 08 March 2011. http://www.finlex.fi/pdf/saadkaan/E0010101.PDF

Parliament of Latvia. 1992. "Republic of Latvia: Law on the Protection of the Body of Deceased Human Beings and the Use of Human Tissues and Organs in Medicine." Accessed 08 March 2011. http://www.vvc.gov.lv/advantagecms/LV/tulkojumi/dokumenti.html?currentPage=11

Parliament of Norway. 2008. "Kingdom of Norway: Act No 44 on Medical and Health Research." Accessed 08 March 2011. http://www.ub.uio.no/ujur/ulovdata/lov-20080620-044-eng.pdf

Parliament of the United Kingdom. 2004. "The United Kingdom: The Human Tissue Act." Accessed 08 March 2011. http://www.legislation.gov.uk/ukpga/2004/30/contents?view=plain

Porteri, C., and P. Borry. 2008. "A Proposal for a Model of Informed Consent for the Collection, Storage and Use of Biological Materials for Research Purposes." *Patient Education and Counseling* 71: 136–42.

Petrini, C. 2010. " 'Broad' Consent, Exceptions to Consent and the Question of Using Biological Samples for Research Purposes Different from the Initial Collection Purpose." *Social Science & Medicine* 70: 217–20.

Price, D. 2005. "The Human Tissue Act 2004." *Modern Law Review* 68 (5): 798–821.

Salvaterra, E., Lecchi, L., Giovanelli, S., et al. 2008. "Banking Together: A Unified Model of Informed Consent for Biobanking." *EMBO Report* 9 (4): 307–13.

Seimas of the Republic of Lithuania. 2000. "Republic of Lithuania: Law on Ethics of Biomedical Research." Accessed 08 March 2011. http://www3.lrs.lt/pls/inter3/dokpaieska.showdoc_e?p_id=148740

Stegmayr, B., and K. Asplund. 2002. "Informed Consent for Genetic Research on Blood Stored for More than a Decade: A Population Based Study." *British Medical Journal* 325: 634–35.

Swiss Parliament. 2009. "Swiss Confederation: The Draft of Swiss Federal Law on Human Research." Accessed 08 March 2011. Available in German at: http://www.bag.admin.ch/themen/medizin/00701/00702/07558/index.html?lang=de

Tarini, B. A., W. Burke, C. R. Scott et al. 2008. "Waiving Informed Consent in Newborn Screening Research: Balancing Social Value and Respect." *American Journal of Medical Genetics Part C (Seminars in Medical Genetics)* 148C: 23–30.

UK Biobank. 2010. "Information Leaflet." Accessed 08 March 2011. http://www.ukbiobank.ac.uk/docs/BIOINFOBK14920410.pdf

UNESCO. 2008. "Report of the International Bioethics Committee of UNESCO (IBC) ON CONSENT." Accessed 08 March 2011. http://unesdoc.unesco.org/images/0017/001781/178124e.pdf

Uranga, A. M., M. C. Martín-Arribas, J. H. di Donato et al. 2005. "Outstanding Legal and Ethical Issues on Biobanks: An Overview on the Regulations of Member States of the EuroBioBank Project." Accessed 08 March 2011. http://www.eurobiobank.org/en/intranet/workflow/uploadDir/PDFmarcadoresEUROBIOBANK-ING.pdf

van Veen, E. B. 2006. "Human Tissue Bank Regulations." *Nature Biotechnology* 24: 496–97.

van Veen, E. B., P. H. J. Riegman, W. N. M. Dinjens et al. 2006. "TuBaFrost 3: Regulatory and Ethical Issues on the Exchange of Residual Tissue for Research Across Europe." *European Journal of Cancer* 42: 2914–23.

Vermeulen, E., M. K. Schmidt, N. K. Aaronson et al. 2009. "Obtaining 'Fresh' Consent for Genetic Research with Biological Samples Archived 10 Years Ago." *European Journal of Cancer* 45: 1168–74.

Wendler, D. 2006. "One Time General Consent for Research on Biological Samples." *British Medical Journal* 332: 544–47.

World Medical Association. 2008. "Declaration of Helsinki – Ethical Principles for Medical Research Involving Human Subjects (Helsinki: 1964; amended Tokyo: 1975; Venice: 1983; Hong Kong: 1989; Republic of South Africa: 1996; Edinburgh: 2000; Washington: 2002; Tokyo: 2004; Seoul: 2008)." Accessed 08 March 2011. http://www.wma.net/en/30publications/10policies/b3/index.html

Chapter 8
Reconsidering Consent and Biobanking

Emma Bullock and Heather Widdows

8.1 Introduction

Increasingly the effectiveness of informed consent as an ethical tool has been questioned both in therapeutic practice and medical research (Katz 1984; O'Neill 2002; Manson and O'Neill 2007). This is particularly the case when genetic information is at stake and in large scale population studies. Accordingly the use of informed consent as an ethical guarantor in biobanking has been questioned (Casado Da Rocha and Seoane 2008).[1] We focus on two difficulties of relying on informed consent in the practice of biobanking. First, an epistemological problem in relation to the information available to the biobank researcher at the time informed consent is sought from the potential donor. Second, the concern that informed consent only respects the rights and interests of the individual from whom samples are taken and does not take into account the rights and interests of third parties.

In order to address these issues we consider a rethought theory of informed consent and alternative ethical models which can be used to replace or supplement it. First, we present and analyse a recent reformulation of the doctrine as a form of waiver (Manson and O'Neill 2007). We argue that although the model is able to bypass the first epistemological difficulty, it fails to meet the second individualist concern. In order to address the second concern – that of respecting the rights and interests of interested third parties – we consider group models. Taken together we argue that the difficulties of obtaining consent to procure and store tissue or information in biobanks can, in principle, be resolved, by both rethinking informed consent and by introducing supplementary group models.

[1]The focus of our paper is on informed consent as an inadequate *feature* of biobank regulation; other ethical guarantees, such as data protection, will not be explored here.

E. Bullock (✉)
Department of Philosophy, University of Birmingham, Birmingham, England
e-mail: ECB767@bham.ac.uk

C. Lenk et al. (eds.), *Biobanks and Tissue Research*, The International Library of Ethics, Law and Technology 8, DOI 10.1007/978-94-007-1673-5_8,

8.2 The Nature of Informed Consent

Medical researchers are required to obtain the consent of their potential subjects before prospective participants can permissibly contribute to the proposed medical research. Whether or not the subject's consent is recognised to be legitimate has been traditionally monitored via the model of *informed consent*. The earliest codification of the doctrine in the Nuremberg Code (1949) states that in order for the potential subject to legitimately consent to partake in the proposed research she must be fully informed of:

1. the purpose of the research
2. the method(s) used, and
3. the likely risks, immediate and projected.[2]

In order to accept consent, scientific researchers must ensure that the potential participants have been given the information desired in conditions (1)–(3). The most recent formulation of the doctrine in the Declaration of Helsinki (2008) expands upon these requirements stating that in order for the potential subject to consent to partake in medical research, she must also be informed of:

4. the sources of funding
5. possible conflicts of interest
6. the intended benefits of the research
7. any other relevant aspects of the study[3]
8. the subject's right to withdraw consent.[4]

The Declaration is thus more demanding about the information that must be communicated to the potential research subject in order for her consent to be fully informed, by requiring *explicit* documented consent procedures and the communication of *specific* information regarding the proposed research (Manson and O'Neill 2007, 8).

[2]"... there should be made known to [the subject] the nature, duration, and purpose of the experiment; the method and means by which it is to be conducted; all inconveniences and hazards reasonable to be expected; and the effects upon his health or person which may possibly come from his participation in the experiment" (Nuremberg Code 1949).

[3]The "relevant aspects" required by condition (7) must be restricted to the broad aims of the study, since the requirement that the potential participant be informed of *any* information relevant to the study is too strong – for instance, it would be impracticable to require that each participant understood the intricacies of the scientific methods used to carry out the research. In this vein, UK Biobank requires only that the potential participant is informed of the broad aims, purpose and nature of the research, reasonably limiting condition (7).

[4]The potential subject must be "adequately informed of the aims, methods, sources of funding, any possible conflicts of interest, institutional affiliations of the researcher, the anticipated benefits and potential risks of the study and the discomfort it may entail, and any other relevant aspects of the study. The potential subject must be informed of the right to refuse to participate at any time without reprisal" (Declaration of Helsinki 2008).

The ethical justification most frequently presented for informed consent procedures is the importance of respecting individual autonomy; the stringent requirements of the doctrine being claimed to be justified by the ethical desirability of respecting the subject's autonomy (Appelbaum et al. 1987; Dworkin 1988; Beauchamp and Childress 1989; Allen and McNamara 2009). Manson and O'Neill have argued that in fact the increasing stringency of conditions imposed by recent codifications of the doctrine is premised on the underlying ethical goal of respecting individual autonomy. As they note: "[t]he reason most commonly given for the expansion, entrenchment and elaboration of informed consent requirements is that they are needed to secure respect for individual autonomy" (Manson and O'Neill 2007, 185). The conditions presented in (1)–(8) are thus defended as *ethical* constraints on the seeking of consent in medical research since the demand to fully inform potential research subjects is a means of respecting individual autonomy. Respect for the autonomous individual thus demands that a research subject's consent is offered on the basis of being informed of the various projected consequences of their participation in the research in accordance with conditions (1)–(8). With regard to biobanking, the doctrine thus requires that potential donors offer explicit and specific informed consent in order that their autonomous decisions are respected.

8.3 Informed Consent and Biobanks

Unfortunately, the traditional doctrine of informed consent is a problematic model for monitoring the legitimacy of consent in biobanks. To see the difficulty of obtaining informed consent in the context of biobanking recall the demands presented in (1)–(8). Put together, these conditions require that the potential research subject be fully informed of the intended use and consequences of her tissue donation or information, and of her right to withdraw consent at any time.

However, these conditions are unsuitable for obtaining the consent of prospective biobank participants for two reasons. The first difficulty is that conditions (1)–(7) are epistemologically opaque to biobank researchers; since the purpose of the biobank is to provide a database of human tissue and medical information which can be accessed by *future* researchers the nature of the medical research utilising this collection will be largely unknown at the time of seeking consent. Although some of the particular research projects might be known when the research subject's data is collected: "the resource will also be used for research that is not yet imagined – so the kind of detailed information about the purpose, methods, risks and benefits usually considered essential cannot be given" (Allen and McNamara 2009, 3). Thus, neither (1) the purpose of the research, (2) the methods to be used, (3) the possible risks, (4) the sources of funding, (5) possible conflicts of interest, (6) the intended benefits of the research, nor indeed (7) relevant aspects of the future study, can be communicated to the potential research subject. From this it follows that the potential subject interested in donating her sample or information to a biobank cannot be sufficiently informed in order for her to offer legitimate consent.[5]

Crucially, according to the informed consent model, if the patient is unable to offer fully informed consent then her autonomy will not be respected.[6] As Hansson has noted, since the future use of the data stored in biobanks cannot be predicted: "[t]he legitimacy of the informed consent depreciates as time proceeds and science makes progress" (Hansson 2005, 415). Accordingly, the patient is unable to give autonomous consent since she lacks the requisite information to make a fully informed decision. Indeed: "because it is impossible for the donor to make an informed choice about the risks and benefits of unspecified future research protocols, such permission should never be called informed consent" (Winickoff and Winickoff 2003). As a result, the traditional model of informed consent is an unsuitable requirement for obtaining consent from prospective participants in biobanks – one cannot be fully informed – therefore one cannot autonomously consent.

A second failing of the model of informed consent in relation to biobanking is that it does not take into consideration the risks and benefits of proposed or future research posed to third-parties. Conditions (3) and (6) which require that the potential participant is informed of the likely risks and benefits of the research are inappropriate conditions for consent if the risks and benefits do not affect the donor, but will instead positively or negatively impact non-consenting third-parties. As Rothstein correctly notes, the risks of biobank research: "go beyond the individual human subjects to population groups with which the subject is associated as well as the general public" (Rothstein 2005, 90). Such risks include the discovery of information that will affect the health of the subject's genetic relations and the possible discriminatory use of the information discovered against certain cultural, geographical and age-related groups (Allen and McNamara 2009). Indeed, unlike other medical information, genetic information provides information about others in addition to the individual from whom samples were taken. For instance, genetic tests on one family member reveal information about the genetic status of other family members; if an individual tests positive for a genetic condition, such

[5]Notably, the subject *is able* to be informed of condition (8) – her right to withdraw her consent at a later time. For instance, UK Biobank grants participants the "right to withdraw": "at any time and without having to explain why and without penalty" (UK Biobank 2007, 6). Options are available with regard to the extent of the withdrawal, from ceasing contact to requiring that their data is put "beyond further use". However, a problem for condition (8) is that it is not possible for the donor to fully withdraw her consent if that is taken to mean removing all record of ever having been involved (technically not possible) nor can one retrospectively retrieve samples and data from completed studies or subsequent studies which build on these studies. Nonetheless, given our focus in this paper on the difficulties of *obtaining* informed consent, further discussion of problems surrounding condition (8) will be left aside.

[6]In addition to disclosure of information, elements of informed consent include voluntariness and competence (Grisso and Appelbaum 1998, 8–10). The claim that a lack of information fails to respect patient autonomy is premised on thought that even if the patient competently and voluntarily consents she cannot autonomously give her consent if she does not know what it is she is consenting to.

as Huntington's disease, or as a carrier of the BRCA1 or BRCA2 (indicators for breast cancer) this information has relevance for family members (consanguineous relations may wish to be tested themselves, and sexual partners may desire the information when making reproductive decisions).[7] This is even more the case when we consider research on indigenous people, as genetic samples taken from an individual in a relatively homogenous ethnic group (for example an indigenous group) gives information about other members of the group. In the case of biobanks there is less risk to the group qua group – as many groups, for instance UK Biobank – are not homogenous and therefore not vulnerable in the same way as a genetically homogenous group (such as an indigenous group). Nonetheless the possibility that related persons (both now and in the future) may be identified from research means that their interests should also be considered. Moreover, as DNA databases expand and genetic makeup and test results are increasingly sought after, the danger of using financially valuable information to identify individuals and related persons increases.

In short, decisions regarding the use of genetic material no longer can be regarded as only of concern to the individuals from whom samples are taken as: "disclosure of genetic information by individual DNA donors also exposes information about others with similar genetic profiles" (Mitchell and Happe 2001, 376). Thus we move necessarily from the individual to the group as the "key feature about genetic information is that it is typically information about a family, or even . . . about a larger community not just about an individual patient" (Brock 2001, 34). Given this, models which only consider the individual are inappropriate in the genetic context. The "genetic self" properly conceived is not an isolated individual but is a "connected self" – connected to consanguineous relations, family groups, ethnic groups, wider communities and in some of the language which surrounds the human genome to "humankind" as a meta-collective (Widdows 2007a, 2009a, b).

When the doctrine of informed consent is used in the context of biobanking, the potential donor is in reality being asked to consent to research which will affect the interests of third-parties. However, since the justification for informed consent is premised upon the ethical importance of respecting individual autonomy, i.e. respect for "the right to make decisions concerning one's own life for oneself without being controlled by others" (Schermer 2002, 23), the individual is not permitted to consent to actions which will significantly affect others who have not been included in the consenting process. The doctrine of informed consent is thus unsuitable in the practice of biobanking where the benefits and risks of the research are likely to affect non-consenting third-parties. As Hansson agrees: "[t]he rule of informed consent in its classical individualistic form is unable to take into account the interests of families and genetic relatives of the patients, individuals who are directly concerned by genetic information" (Hansson 2005, 415). In biobanks this is especially problematic as it is likely to affect future persons – who clearly cannot be included in any form of consent, however presumed. Since the benefits and

[7] For further discussion on this please see previous paper (Widdows 2007b).

risks posed by research involving biobank data extend beyond the interests of the potential participant to present and future third-parties, the demand for informed consent is ethically improper.

A few responses can be made to the epistemological and individualist inadequacies of the traditional model of informed consent with regard to biobanking. Firstly, in order to avoid the epistemological problem informed consent could be sought every time an individual's data was used or reused. Secondly, biobanks could be rejected altogether as an acceptable resource for medical research since they fail to meet the standards required by the doctrine of informed consent. Finally the doctrine of informed consent could be rejected, adapted and/or supplemented by other ethical models in order to bypass both the epistemological problem and the difficulty of respecting communal interests.

We opt for the latter response on the basis that the first two options are undesirable. Firstly, the attempt to enforce informed consent procedures every time an individual's data is used in the future has been rejected as unworkable. This is particularly true in large scale population research, such as biobanks. In such instances to insist on fully informed consent would be both impracticable and unrealistic (Widdows 2009a, 85). Such a requirement would demand returning to the donors for every new study – and potentially for every subsequent study which drew on previous data. To do this would not only be administratively cumbersome but more importantly overly burdensome on the donors to the point of impossibility. As Allen and McNamara have defended: "obtaining consent for each new use is impractical and . . . the burden of compliance is prohibitively expensive" (Allen and McNamara 2009, 3–4).

Secondly, the abolition of biobanking altogether is too drastic, since they provide an invaluable resource for furthering medical science. As Rothstein rightly points out: "[b]iomedical research is heavily supported and greatly valued by society because it advances scientific knowledge and leads to interventions to prevent, treat, and cure human afflictions" (Rothstein 2005, 89). Specifically, the UK Biobank is currently collecting and collating biological data in order to provide a resource to support the future research of diseases such as cancer, heart disease, stroke, diabetes and dementia (Biobank UK 2007). The complete rejection of biobanks would be a cost to medical progress and so must only be thought of as a last resort. Moreover, simply halting biobanks would do nothing to address the ethical challenges presented by the communal and identifying nature of genetic information. While informed consent is particularly problematic in biobanking because of their future orientated status it is problematic for all clinical and research practice in the genetic era. The identifying nature of genetic material means that the rights and interests of third parties can no longer be ignored; hence individual consent would need revision even if biobanks were prohibited. Therefore, rather than rejecting the use of biobanks we need to address the problems with informed consent in all spheres of medicine in the genetic era, including biobanks.

In the next sections we consider a recent attempt to reconceptualise informed consent as a form of "waiver" and the supplementation of individual models of consent with group models. The appropriateness of these models is based on their ability to address some of the failings of the traditional doctrine of informed consent, including both the epistemological difficulty of offering consent for future research and the problem of protecting the rights and interests of third-parties.

8.4 Informed Consent as Waiver

A revised model of informed consent has been presented by Neil Manson and Onora O'Neill (2007). They refer to their theory as "the waiver model" since instead of fully informed consent being viewed as a means of respecting individual autonomy informed consent is treated as a waiver of certain rights and obligations. Manson and O'Neill argue for the waiver model on the basis that respect for individual autonomy is an unsuitable consideration for justifying informed consent procedures, since they correctly point out that it is unclear on both minimal and rational accounts why autonomy ought to be respected (Manson and O'Neill 2007, 20–21).[8] We argue that although the waiver model of consent addresses the epistemological problem, it fails to resolve the difficulty of respecting communal rights and interests. Furthermore, the waiver model highlights a problem with viewing consent as sufficient as an ethical justification for action.

Rather than being justified by the ethical need to respect autonomous decisions, Manson and O'Neill argue that informed consent is justified by the necessity of waiving certain rights and obligations in the medical arena. They develop their argument by noting that a feature of medical research is that it infringes an individual's rights. The waiver model of informed consent is thus seen by Manson and O'Neill as: "a way of justifying action that would otherwise violate important norms, standards or expectations" (Manson and O'Neill 2007, 75). Furthermore, not only does the waiver model explain why informed consent is important from the perspective of the patient, it also illuminates the importance of consent procedures for both doctors and researchers. As Davis points out: "[c]rimes and torts may not be committed if a person properly consents to an interference which would otherwise be unlawful" (Davis 2003, 368). Thus, if the patient is viewed as waiving her rights under the revised model of informed consent then the doctor or researcher is not criminally accountable for actions she undertakes in accordance with the donor's consent.

In order to clarify the rights that may be permissibly waived Manson and O'Neill distinguish between *intrusive* and *invasive* action. Whilst *invasive* action will usually infringe a person's right to bodily integrity, *intrusive* action: "infringes a specific range of liberty rights, often referred to as privacy rights" (Manson and O'Neill 2007, 97). Intrusive action includes the use of biological data in the process of

[8] Since we have already motivated the desirability of seeking alternative models for obtaining consent in the context of biobanking, Manson and O'Neill's impetus will not be explored here in detail.

medical research. According to the theory, the donor may thus consent to the use of her private data in research by waiving some of her privacy rights.

Notably, there is a possibility of viewing the waiver model as a form of broad consent, defined as consent to unspecified future research such that the nature of the research is only generally, or broadly, known. Broad consent alone is unacceptable as is gives donors no say in the way their material is actually used. "One-off" consent or the long-term waiving of a privacy right thus requires that it be fully informed. Not to do so would be to effectively licence researchers to do anything they wish with the material and to grant donors no rights or interests in their material.

In view of the need for adequate information, Manson and O'Neill argue that the waiver model of informed consent must be underpinned by an agential model of communication; informed consent as waiver is only successfully met when the potential research subject and researcher are "sensitive to one another as agents with their own cognitive and practical commitments, and assume one another's adherence to a range of communicative, epistemic and ethical norms" (Manson and O'Neill 2007, 66). A further requirement that the theory of informed consent as waiver must meet is thus that the information demanded in conditions (1)–(8) is communicated in light of a variety of communicative norms. These will include sharing a language, sharing background knowledge of the world, inferential competence and knowledge of each others' commitments and competencies (Manson and O'Neill 2007). With this explicit account of communication in mind, the general purpose, use and risks associated with the future research can be better communicated to the potential donor; the epistemological problem can be partially met by requiring that consent to general future research is informed. The waiver model is thus improved by demanding that certain communicative demands are explicitly met before the potential donor waives her right to privacy, thereby avoiding the epistemological problem raised against the traditional model of informed consent.

Nevertheless, even if the nature of future research can be explicitly communicated to the potential donor, the waiver model fails to succeed in legitimizing biobank research. Firstly, although there seems to be no *epistemological* difficulty for an individual to waive some of her rights to privacy over an extended period of time if she has *enough* information about the nature of the research, it is not clear that consent is the appropriate justification for biobank research. Brownsword (2004, 2009) has argued that it is fallacious to suppose that consent makes an action ethically legitimate. Rather, in order for the waiving of rights to be ethically significant, one needs to discern what *permits* someone to waive their rights and whether this is possible. As Davis accurately observes, a patient's express consent to the waiving of certain human rights "may be insufficient to override the public purposes that justify the illegality of the activity in the first place" (Davis 2003, 368–69). Determining the legitimacy of rights-waiving has been shown to be problematical in British case law. In R. v. Brown (1993) consenting sadomasochists who committed acts of violence upon one another "including genital torture, for the sexual pleasure which it engendered in the giving and receiving of pain" (R. v. Brown 1993, 76) were convicted of assaults occasioning actual bodily harm. An appeal was made against the conviction under the contention that "a person could not guilty be [sic] of assault occasioning actual bodily harm or unlawful wounding in respect of acts carried out in private with the consent of the victim" (R. v. Brown 1993, 76). As part of their appeal they thus proposed that individual consent was sufficient to waive rights to bodily integrity.

However, individual consent was found to be an insufficient justification for the waiving of this right. As Lord Jauncey of Tullichettle noted: "there must be some limitation upon the harm which an individual could consent to receive at the hand of another" (R. v. Brown 1993, 86). The consent of the sadomasochists was found by the court to be insufficient to waive their rights to bodily integrity. As this example shows, before the waiving of human rights can be accepted as being ethically justifiable it is necessary to discern the conditions under which it is morally legitimate. Thus, given that there are limitations on what we can consent to, further restrictions (or additions) need to be in place in order to monitor the procurement, storage and use of data in biobanks on the waiver model. Possible supplements necessary on the waiver model are group models which will be considered in the next section.

Secondly, the waiver model is also unable to address the second problem rallied against the traditional doctrine of informed consent in that no matter how successfully an individual can waive some of her rights this ability does not extend to waiving the rights of third-parties. Brownsword illustrates the ethical inappropriateness of consenting to an action which affects non-consenting parties with an example:

> A acts with the consent of B, and vice versa. A does no wrong to B, and B does no wrong to A. However, if the transaction between A and B violates the rights of C, there is a wrong to C. For example, if A and B agree to paint-spray the slogan "Consent rules OK" on C's car or on C's house, the fact that A and B have so consented is no response to C. C's rights have been violated. (Brownsword 2004, 240)

In this case, the consent of A and B is insufficient to legitimize the action since the result affects the interests of and is unwanted by C. Hence, despite addressing some aspects of the epistemological difficulty, the waiver model fails to explicate the nature of the rights an individual can legitimately waive, both in terms of her own interests, and the rights and interests of others. In light of the limitations of the waiver model of consent, we thus supplement the waiver model with group models below.

8.5 Group Models of Consent

The criticism that informed consent is incapable of addressing the needs of third parties has been the biggest impetus of rethinking informed consent. This has been true particularly with regard to genetics where communal models are principally needed. This is for the two main reasons outlined in Section 8.3; first that genetic research, particularly biobanking research, is often on-going and future orientated. Second and most relevant to this section of the paper, genetic information is identifying not just of the individual from whom the sample was taken but also of related individuals.

These problems with informed consent are compounded in genetic research as not only is individual informed consent an inadequate ethical tool, but confidentiality is also compromised (Husted 1997; Knoppers 1997, 1999; Widdows 2007b, 2009a). Simply put genetic information is identifying and therefore confidentiality can never be guaranteed – no matter how anonymised the data is – as there is always a risk of identification (of the individual, related individuals or the group) when the information is compared with a database. This risk grows as the number of databases grows and thus the possibility of identification increases.[9]

Concerns about the adequacy of informed consent and confidentiality have lead to the "communal turn" in ethics (Knoppers and Chadwick 2005; Widdows 2009a). Ethical frameworks are being sought which will take into account the needs of third parties. This is happening at all levels of clinical and research genetics and the seeking of effective ethical models in biobanking is part of a broader search for better ethical models. For example, in the context of genetic testing family based frameworks have begun to emerge, such as the "family covenant" and the "joint account" model.[10] Models which have arisen in the context of population genetics and biobanking include those of group consent, benefit-sharing, group patenting, trust and conditional gift. These models can be used in conjunction with individual consent (either broad or wavier consent) to produce a satisfactory ethical model in instances where informed consent is no longer sufficient or effective in meeting ethical requirements.

Different models have been used alone or in conjunction to meet different needs. For instance, group consent has emerged as an attempt to respond to the criticisms of the HGDP and the examples of bad practice of patenting of indigenous DNA.[11] Research councils have: "proposed the adoption of 'group consent' as a normative rule governing genomic research to alleviate this ethical blind spot in traditional

---

[9] For further discussion on the problems of confidentiality in the genetic era see Widdows (2007b).

[10] Possible family models suggested in the literature are the "joint account" (Parker and Lucassen 2004) and the "family covenant" (Doukas and Berg 2001). The joint account model sees genetic information as belonging to, and available to, all family members. While there may be cases where information could be withheld (for example in situations where such disclosure would seriously harm the individual) the "default" position would be that genetic information is familial and not individual (Parker 2001). The "family covenant" is perhaps the most discussed model of family consent, where the family and not the individual is the "unit of care". The family covenant is suggested as a model which dictates the manner in which results from genetic tests are to be shared with family members, as such it "offers the individual, family and physician a mechanism to help resolve competing claims for confidentiality and disclosure" (Doukas and Berg 2001, 3). It was intended to pre-empt questions of when and how to disclose potentially distressing information about the genetic status of individuals within families. The thinking behind this approach is that the "bonds that hold families together may not survive such dynamic tension unless there is some framework constructed to allow for balancing of individual and family interests" (Doukas and Berg 2001, 3) These models are discussed in more detail elsewhere (Widdows 2007b).

[11] For instance to prevent the wrongs such as those done to the Hagahai tribe from occurring in the future (Widdows 2009a).

informed consent doctrines" (Mitchell and Happe 2001, 377).[12] Such attempts are becoming standard practice and some kind of group consent in the form of: "prior consultation and communication with these specific communities and populations are emerging as ethical prerequisites" (Knoppers and Chadwick 2005, 76). Benefit sharing has also become more commonly utilised and it has been recommended by the HUGO Ethics Committee that this model be used and benefits agreed before the research begins: "consultation with individuals and communities and their involvement and participation in the research design is a preliminary basis for the future distribution of benefit and may be considered a benefit in itself" (HUGO Ethics Committee 2000, sec. G).

Both group consent and benefit sharing clearly go some way to address the problems of meeting the needs of interested third-parties. However, while there are instances where they have worked well with homogenous groups, for instance indigenous groups, it is less clear how applicable they are for less homogenous groups – for instance, for a group like UK Biobank participants. It is not impossible to imagine some form of group consent in UK Biobank – for instance, one possibility would be to do it via parliament or as a condition of inclusion in the NHS.

For the reasons discussed in Section 8.3, return to research subjects would be impracticable and unrealistic as UK Biobank will recruit 500,000 people aged 40–69. It will take physical samples, ask lifestyle questions, and link this information to health-relevant records. To require frequent return to the donors for every new study would be impossible and overly burdensome. In a model of conditional gift broad consent is limited by conditions on all subsequent research using the material. The donors' conditions are set out in the initial "broad consent" – i.e. "my samples can only be used for research on X, Y and Z". In the trust model broad consent is limited by the trust framework: i.e. the institution must respect the terms of the consent. The donor gives broad consent on trust that their material will be used as promised in the original consent and that this will be ensured by the on-going ethical and governance mechanism. The trust model is less rigid in its conditions than the conditional gift model in that X, Y and Z are not specified in detail. However, the trust model does require that the samples are used in accord with the original conditions (however broad these may be).

In the trust models additional ethics and governance mechanisms, are introduced to supplement consent and to ensure that the samples are used in accord with the expectation of the donor and with the conditions on the original consent. In the case of UK Biobank the consent is "broad", but the waiver model could also be used, and we would suggest that it should be. In the trust model: "when a person agrees to donate tissue, the recipient has a responsibility to serve as a trustee, or steward, of the tissue in order to ensure protection of the contribution" (Winickoff and Winickoff 2003, 1182). The open-ended nature of this consent is made clear to participants from the outset. It is overtly not "fully informed", indeed this would

[12]For example see the Nuffield Council on Bioethics document *The ethics of research related to healthcare in developing countries*.

be impossible given that the nature of the research is as yet unknown (UK Biobank 2007, 6). Essentially the trust model is "broad consent" – or as we would suggest waiver consent – with additional safeguards and governance mechanisms. However, while it has some parallels with a simple broad consent model it should not be confused with it as the additional mechanisms are absolutely fundamental. Possible safeguards and additional mechanisms suggested in the literature: "include membership on the trust's IRB, membership on a donor committee that has veto power over particular projects, and election of a donor to serve on the board of trustees. Research applications could be evaluated by the trustees according to a set of criteria that would ensure public benefit" (Winickoff and Winickoff 2003, 1182–83).

In the case of UK Biobank, participants agree that their samples can be used in ways which fit the stated purpose of UK Biobank; that is, to "build a major resource that can support a diverse range of research intended to improve the prevention, diagnosis, and treatment of illness and the promotion of health throughout society" (UK Biobank 2007, 3). When asked to consent participants are asked on the basis that participation is "an opportunity to contribute to a resource that may, in the long term, help enhance other people's health" (UK Biobank 2007, 5). While the stated aims are broad, this is not simple broad consent; research is limited by the consent. First research must be done in accord with UK Biobank's stated aims and second this is monitored and ensured by the on-going ethics and governance mechanisms.

First then, the participants material can *only* be used in accordance with the stated aim: that: "the resource is being used in the public interest" (UK Biobank 2007, 13). Determining the public interest in detail is potentially problematic, however, broad principles are not difficult to determine. For instance, research which posed a public health risk would be unacceptable – whatever the expected scientific benefit – as perhaps would be research leading to developments which the donor group would not have access to (for instance, due to a high cost).[13] Public interest is therefore central to this model: participants give their data to the biobank to serve the public interest and the biobank conducts research in the public interest. Group models thus meet the second objection by explicitly taking account of the interests of third parties – both now and in the future.

Second, the additional ethics and governance mechanisms which supplement the individual consent (whether waiver or broad) protects participants interests by transforming consent from a "one off" act to an ongoing process. Thus the ethical focus is not only at the beginning of the research (at the point of consent) but, by incorporating additional ethical and governance mechanisms, occurs throughout the lifetime of the project or the research. In the case of UK Biobank the additional ethical and governance mechanisms are the Ethics and Governance Framework (EGF), which sets out the aims of UK Biobank, with which UK Biobank is bound to comply, and the Ethics and Governance Council (EGC). The EGC's remit is:

[13] It is important to note that while there is overlap the public good and the scientific good are not equivalent.

> acting as an independent guardian of the *Ethics and Governance Framework* and advising the Board on its revision; monitoring and reporting publicly on the conformity of the UK Biobank project within this Framework; and advising more generally on the interests of participants and the general public in relation to UK Biobank. (UK Biobank 2007, 15)

In the case of UK Biobank it is the EGC's responsibility to fulfill this stewardship role and ensure that the rights and interests of the donors and the wider public are protected and safeguarded. In this model consent remains part of the process – functioning at entry and exit (participants have the right to withdraw) – but they are protected not because of the consent, which is broad or waived but because: "the Ethics and Governance Council will keep use of the resource under review in order to advise on conformance with this Framework . . . to assure itself, and others, that the resource is being used in the public interest" (UK Biobank 2007, 13). Therefore the point of consent is only one factor in protecting the rights and interests of participants (and the public interest). More important are the supplementary ethics and governance mechanisms – the EGF and EGC. Accordingly group models can supplement consent and ethical frameworks can be developed which respect the rights and interests of donors into the future and of third parties, thereby meeting the initial problem of respecting communal rights.

8.6 Conclusions

The traditional doctrine of informed consent is an unsuitable model for governing the ethics of procuring consent to store tissue or information in biobanks. This is due to two problematic features of the doctrine; firstly there is an epistemological difficulty given the future directedness and indefinite nature of the research to be carried out meaning that the requirement of the doctrine that the subject be fully informed cannot be met. Secondly, the doctrine of informed consent is too individualistic to account for the positive or negative impact such research might have on interested third-parties without their consent.

In response to these problems we have explored a number of possible alternatives. Firstly we have looked at a recent revision of the doctrine – informed consent as waiver. Unlike the traditional model, the waiver model of consent is premised upon the legal necessity of waiving certain privacy rights in the realm of medical research. Although the waiver model improved upon the traditional account by circumventing the epistemological problem with an agential account of communication, the model remained too individualistic to fully address the initial demand that seeking consent for participation in biobanks needs to be supplemented by communal concerns. These limitations can be addressed if additional ethics and governance mechanisms are used to either replace or supplement individual consent. To this end we explored group models and suggested that taken together these alternatives do address some of the problematic features of informed consent and provide some suggestions as to how to establish more effective ethical models for biobanking.

References

Allen, J., and B. McNamara. 2009. "Reconsidering the Value of Consent in Biobank Research." *Bioeth* 25 (3): 155–66.

Appelbaum, P. S., C. W. Lidz, and A. Meisel. 1987. *Informed Consent: Legal Theory and Clinical Practice*. New York: Oxford University Press.

Beauchamp, T. L., and J. F. Childress. 1989. *Principles of Biomedical Ethics*. 3rd edn. New York: Oxford University Press.

Biobank UK. 2007. "Further Information Leaflet." Accessed February 25, 2011. http://www.ukbiobank.ac.uk/docs/urtherinfoleaflet241007.pdf.

Brock, D. W. 2001. "Genetics and Confidentiality." *The American Journal of Bioethics* 1 (3): 34–35.

Brownsword, R. 2004. "The Cult of Consent: Fixation and Fallacy." *KCLJ* 15: 223–51.

Brownsword, R. 2009. "Rights, Responsibility and Stewardship: Beyond Consent." In *The Governance of Genetic Information: Who Decides?*, edited by H. Widdows and C. Mullen, 99–125. Cambridge: Cambridge University Press.

Casado Da Rocha, A., and Seoane, J. A. 2008. "Alternative Consent Models for Biobanks: The New Spanish Law on Biomedical Research." *Bioethics* 22 (8): 440–47.

Davis, H. 2003. *Human Rights and Civil Liberties*. Devon: Willan Publishing.

Declaration of Helsinki. 2008. "Section 24." Accessed February 25, 2011. http://www.wma.net/en/30publications/10policies/b3/index.html.

Doukas, D. J., and J. W. Berg. 2001. "The Family Covenant and Genetic Testing." *The American Journal of Bioethics* 1 (3): 2–16.

Dworkin, G. 1988. *The Theory and Practice of Autonomy*. Cambridge: Cambridge University Press.

Grisso, T., and P. S. Appelbaum. 1998. *Assessing Competence to Consent to Treatment*. Oxford: Oxford University Press.

HUGO (Human Genome Organisation) Ethics Committee. 2000. "Statement on Benefit Sharing." Accessed March 1, 2011. http://mbrdb.nibio.go.jp/kiban01/downloadEN/HUGO/Statement_on_Benefit_Sharing.htm

Hansson, M. G. 2005. "Building on Relationships of Trust in Biobank Research." *Journal of Medical Ethics* 31: 415–18.

Husted, J. 1997. "Autonomy and a Right Not to Know." In *The Right to Know and the Right Not to Know*, edited by R. Chadwick et al, 55–68. Aldershot: Ashgate.

Katz, J. 1984. *The Silent World of Doctor and Patient*. The John Hopkins University Press: London.

Knoppers, B. M. (Ed.). 1997. *Human DNA: Law and Policy*. The Hague: Kluwer Law International.

Knoppers, B. M. 1999. "Who Should Have Access to Genetic Information." In *The Genetic Revolution and Human Rights*, edited by J. Burley, 54–60. Oxford: Oxford University Press.

Knoppers, B. M., and R. Chadwick. 2005. "Human Genetic Research: Emerging Trends in Ethics." *Nature* 6: 75–79.

Manson, N. C., and O. O'Neill. 2007. *Rethinking Informed Consent in Bioethics*. Cambridge: Cambridge University Press.

Mitchell, G. R., and K. Happe. 2001. "Informed Consent after the Human Genome Project." *Rhetoric and Public Affairs* 4 (3): 375–406.

Nuremberg Code. 1949. "Directives for Human Experimentation." Accessed February 25, 2011. http://ohsr.od.nih.gov/guidelines/nuremberg.html

O'Neill, O. 2002. *Autonomy and Trust in Bioethics*. Cambridge: Cambridge University Press.

Parker, M 2001. "Confidentiality in Genetic Testing." *The American Journal of Bioethics* 1 (3): 21–22.

Parker, M., and A. M. Lucassen. 2004. "Genetic Information: A Joint Account?" *British Medical Journal* 329: 165–67.

R. v. Brown and other appeals. 1993. 2 All ER 75.

Rothstein, M. A. 2005. "Expanding the Ethical Analysis of Biobanks." *Journal of Law, Medicine and Ethics* 33: 89–101.

Schermer, M. 2002. *The Different Faces of Autonomy: Patient Autonomy in Ethical Theory and Hospital Practice*. London: Kluwer Academic Publishers.

UK Biobank. 2007. *Ethics and Governance Framework*. Accessed February 25, 2011. http://www.ukbiobank.ac.uk/docs/20071011_EGF_Version_3_1_0October_2007withTOR.pdf

Widdows, H. 2007a. "Conceptualising the Self in the Genetic Era." *Health Care Analysis* 15: 5–12.

Widdows, H 2007b. "Reconceptualising Genetics: Challenges to Traditional Medical Ethics." In *Ethics and Law of Intellectual Property: Current Problems in Politics, Science and Technology*, edited by C. Lenk, N. Hoppe, and R. Andomo, 159–74. Aldershot: Ashgate.

Widdows, H 2009a. "Constructing Communal Models of Governance." In *The Governance of Genetic Information: Who Decides?* edited by H. Widdows and C. Mullen, 75–98. Cambridge: Cambridge University Press.

Widdows, H 2009b. "Between the Individual and the Community: The Impact of Genetics on Ethical Models." *New Genetics and Society* 28 (2): 173–88.

Winickoff, D. E 2003. "Governing Population Genomics." *Jurimetrics* 43: 187–228.

Winickoff, D. E., and R. N. Winickoff. 2003. "The Charitable Trust as a Model for Genomic Biobanks." *The New England Journal of Medicine* 349: 1180–84.

Chapter 9
What's Wrong with Forensic Uses of Biobanks?

Claudio Tamburrini

Privacy is increasingly becoming a more and more serious concern in the context of biobanking. For that reason, the anonymization and pseudonymization of samples donors' data have lately attracted much research work. However, information loss is an unfortunate consequence of efficient anonymization. This is particularly evident in the area of forensic uses of biobanks data. Usually, the question is presented in terms of choosing between, on the one hand, the promise of a very effective instrument to fight criminality and, on the other hand, the threat of granting the State access to confidential information about our genetic constitution. More concretely, opponents of this practice object that forensic data bases (i) discriminate against certain social groups, particularly when the data are kept even after the suspect has been dismissed from the investigation or acquitted in trial; (ii) lead to miscarriages of justice, as it is shown by some cases where innocents were found guilty because of errors committed in genetic data analysis; (iii) can be misused by governments to control citizens through information storage that might be used against them in the future; (iv) violate donors' privacy, particularly as genetic data banks imply that confidential information about donors' – as well as their relatives' – propensity to develop certain diseases is collected and put at researchers' – or State authorities' – disposal. Finally, it is also argued that all these problematic aspects of forensic biobanking, as they were expressed in the objections above, (v) can be conducive to the discredit of genetic biobanks in general, thus weakening people's willingness to contribute their samples to the repositories.

Among the objections listed above, it is particularly the right to privacy that is most often invoked in the debate on forensic biobanks. The objection from privacy is generally seen as both relevant and strong, even when donors explicitly might have waived this right.

There is however a remarkable lack of conceptual accuracy regarding both the nature and the content of this right. Obviously, this has a direct bearing on how the other objections should be judged. So, before we can decide whether or not it might

C. Tamburrini (✉)
Centre for Healthcare Ethics, Stockholm University, Stockholm, Sweden
e-mail: claudio.tamburrini@philosophy.su.se

C. Lenk et al. (eds.), *Biobanks and Tissue Research*, The International Library of Ethics, Law and Technology 8, DOI 10.1007/978-94-007-1673-5_9,

be justified to override donors' right to privacy in certain occasions (for instance, to solve a serious crime), we need to ascertain (i) what is generally meant by the right to privacy, and (ii) which strength can reasonably be attached to this demand. This conceptual analysis is a prerequisite for adopting a well-grounded stance on all the objections mentioned above.

Almost equally important is to clarify which kind of biobanks will be referred to in this article. The biological samples that might be used for forensic purposes can be gathered in different ways. Most biobanks are created with the explicit aim of serving clinical and research purposes. Others instead are implemented within the legal system to be applied to forensic matters. A clear though succinct characterization of biobanks is therefore needed to provide stringency to our discussion.

The structure of this article is as follows. In the first section, I start with a brief, introductory account of the different ways of making sample donors' anonymized and/or pseudonymized. In Section 9.2, I distinguish between different kinds of biobanks mainly on grounds of whose samples are collected and the areas of application. In Section 9.3, I set out to characterize the notion of privacy. Section 9.4 is centred on different levels of strength that might be attributed to a particular right, properly illustrated by examples gathered from health care practice. In the remaining sections, objections (i) through (v) as they have been formulated above are discussed.

Finally, a remark on what is not included in this article. The efficiency of forensic data basis for clearing up crime is still much questioned. However, whatever the state of the art might be in that regard, it can hardly be questioned that – when properly developed – DNA-techniques will greatly contribute to solving crimes. Accordingly, I will not discuss the efficiency issue in this article, but will instead take for granted that forensic DNA-techniques work – or will work in the future – as a crime solving device. No unnecessary boldness is implied in such an assumption.

9.1 Anonymization and Pseudonymization in Biobanking

Collecting biological samples is becoming an increasingly important tool both in clinical practice and research activities. Accordingly, the number of biobanks has exploded during the last years. In order to improve research into human diseases and their genetic and physiological causes, we need to keep this trend by securing a continuous flow of donations to biobanks. The expansion of biobanks has been followed by increased concern about donors' rights, particularly their right to privacy. Different anonymization procedures have therefore been designed to keep the identity of the donors confidential. There is no commonly-agreed terminology regarding the different anonymization procedures.[1] However, the terminology used

[1]Terminology discrepancies are particularly salient between common law and continental traditions. For a thorough account of these discrepancies, see Elger and Caplan (2006).

in many European documents to refer to these different levels of anonymization can be summed up as follows: (a) unlinked anonymized; (b) linked anonymized (or pseudonymized) and (c) coded.[2]

In the context of biobanks, the term "unlinked anonymized" means that a certain biological material is stored together with other related data, such as the type of disease, the medical treatment, the donor's age, etc., but all other information that might allow the identification of the donor is *irreversibly* erased. When the stripping of the information linking the samples to a particular donor is made reversible (that is, it can be traced back to the donor), then we speak of "linked anonymized" or "pseudonymized" samples. Here the identification of the donor is made possible by using a code to which researchers or other users of the biobank samples do not have access. Finally, "coded samples" are stored according to the same procedures as linked, reversibly anonymized samples, with the difference that researchers and users do have access to the code.

9.2 Different Kinds of Biobanks

Different types of forensic biobanks might be distinguished according to *how the samples are gathered* and, partly related to this, *which kind of control donors have over their samples*. Although the following classification is not exhaustive, I think at least three different kinds of forensic biobanks could be outlined:

(a) Special forensic data bases which are totally separated from medical biobanks, either already existing or in preparation. The biological material is compulsorily gathered from suspects and/or convicted felons.
(b) Forensic data bases created by enlarging the area of application of existing medical biobanks. The biobank population is conformed by all those who voluntarily agree to contribute with a sample.
(c) As (b) above, with the difference that the data bases now consists of all the genetic information from newborns gathered at birth for medical research purposes, with or without any explicit statement that would preclude forensic uses. Samples gathering is also voluntary, although mediated by the consent from parents or legal representatives on behalf of their children.

In this article I will primarily refer to alternatives (b) and (c). The reason for this limitation is that, when criminal conduct is involved, the exercise of rights usually

[2]Some of the documents in which this terminology can be found are, for instance, CDBI (2006), Draft explanatory memorandum to the draft recommendation on research on biological materials of human origin. Strasbourg, France: Council of Europe Steering Committee on Bioethics; COE (2006), Recommendation Rec (2006)4 of the Committee of Ministers to member states on research on biological materials of human origin. Strasbourg, France: Council of Europe.

recognized to ordinary citizens is suspended.[3] So, the central question of this article might now be formulated in the following terms: *Would it be justified to hand over samples belonging in biobanks and/or genetic new-borns banks for forensic purposes, even without the consent of the donors or their legal representatives?*

9.3 The Notion of Privacy

What is meant by the notion of privacy? Although this is a recurrent argument in the medical ethics debate, it is far from clear what it stands for. In a general sense, privacy refers to protecting individuals' physical and psychological integrity from illegitimate intrusion. As such, privacy is valuable because it protects what people consider to be important in life, "such as the intimate sphere or the conditions for autonomous judgement."[4] Thus, the right to privacy is often expressed in connection with a concern for individual autonomy, understood as a right of individuals to be respected as agents fully capable of making and carrying out their own decisions. In DeCew's words,

> Privacy acts as a shield to protect us in various ways, and its value lies in the freedom and independence it provides for us. Privacy shields us not only from interference and pressures that preclude self-expression and the development of relationships, but also from intrusions and pressures arising from others' access to our persons and the details about us. Threats of information leaks as well as threats of control over our bodies, our activities, and our power to make our own choices give rise to fears that we are being scrutinized, judged, ridiculed, pressured, coerced, or otherwise taken advantage of by others.[5]

Among the issues which are considered of importance for personal autonomy we find, for instance, which research projects the samples may be used for, the possibility of opting out from a biobank after one has contributed with a sample and the prerogative of deciding oneself whether or not to disclose personal, sensitive information to third parties. More specifically, as genetic data banks store data on donors' – as well as their relatives' – propensity to develop certain

---

[3]Something similar applies to the notion of consent, as a criminal can be sent to prison in spite of her not accepting the punishment. However, this might be different in other areas in which consent also is a relevant notion. Think, for instance, of a practice that implied that imprisoned criminals are compulsory made to donate biological samples to a biobanking or, where capital punishment still is enforced, to donate their organs without their consent. It is therefore far from obvious that privacy, consent and other basic rights can without problem be set aside, even in criminal contexts. As this aspect exceeds the scope of this article, it will not be further discussed.

[4]Beckman (2005, quotation on p. 98). This reference has been gathered from Teetzel (2009, quotation on p. 43).

[5]DeCew (1999, quotation on pp. 249). This reference has been gathered from Teetzel (2009, quotation on pp. 43–44).

diseases, it is feared that personal autonomy might be jeopardised if this information is disclosed to unauthorised parties (for instance, insurance companies and employers).

This characterization can also be found both in the European legislation and in the European Court praxis. In a leading case that referred to Art. 8 of the European Convention of Human Rights, the European Court stated that the concept of a private life also

> includes a person's physical and psychological integrity; the guarantee afforded by Art. 8 of the Convention is primarily intended to ensure the development, without outside interference, of the personality of each individual in his relations with other human beings. There is therefore a zone of interaction of a person with others, even in a public context, which may fall within the scope of "private life".[6]

So, the conceptual analysis of the notion of privacy points out a *wider sense* (privacy as the protection of one's physical and psychological integrity) as well as a *narrower one* (privacy as the protection of confidential information). As we will see later on in this article, both are relevant for the issue of the forensic uses of stored biological samples.

9.4 Which Kind of Right Is the Right to Privacy?

Let us now characterize the notion of a right. What does it mean to say that a person is the bearer of a particular right? This could be understood as a demand on others that implies that they have to perform something concrete, as for instance when we say that individuals in a welfare state have a right to free health care. Usually we refer to this kind of rights as positive rights.

A right can also be seen, not as creating a positive obligation on others to provide us with certain things or services, but instead as a constraint imposed on their conduct. The bearer of this kind of negative rights is therefore entitled to be left alone within her private sphere, unless s/he accepts otherwise. The right to privacy should most reasonably be seen as a negative right, as the alternative of requesting others to take active steps to further our privacy seems too demanding.

The right to privacy is therefore more reasonably seen as a constraint on the actions of others intended to safeguard (at least part of) our physical and psychological integrity. But which strength does this right have?

To say that the right of privacy is just one among a number of prima-facie prerogatives recognised to individuals would be too weak an interpretation. Of course, this is not to say that the right to privacy is *never* seen as exactly that weak in medical ethics. At least in one of its sub areas – sport medical ethics – athletes' privacy is not much worth. As we know, athletes are constantly required to inform sporting authorities about their whereabouts in order to make themselves available for

[6] In Von Hannover v Germany (2004).

unannounced doping tests. And, if that is required to fight cheating in sport competitions, the hematocrite level of an athlete, together with other sensitive medical information, can be delivered to the public and the media by sport medical doctors with practically very few restraints. These examples notwithstanding, I think that in the wider health care system, privacy enjoys at least some moral and legal weight.

How much weight? Well, certainly not the weight that might be attributed to an absolute right, never to be trumped by another competing claim. Think, for instance, of the widely recognised right of legally competent patients to refuse medical treatment they, for some reason, don't want to undergo. In that case, there is no competing claim that might in the end turn the balance in favour of submitting the patient to compulsory treatment. Patients' right to refuse treatment is therefore an *absolute, non-overridable* one, in a way that the right of privacy is not.

Rather, privacy is generally considered as an absolute right which nonetheless can be overridden by other competing claims. This is also the way in which consent agreements in biobanks usually are cast. Privacy and donors' right not to be a part of a project they for some reason reject are often recognised, though conditionally. These patients prerogatives can in some circumstances be set aside, for instance when that "might be required to further a particularly valuable research project". This is not only a truthful characterization of biobanking practice; it is also the most reasonable interpretation of the right to privacy in bio-ethical contexts: an *absolute though overridable right*. Given the utility derived from clinical and research uses of biobanks samples, any other alternative implies that valuable social goals might be jeopardised.

9.5 Forensic Uses of Biobanks

If donors' right to privacy is most reasonably understood as an overridable demand on researchers and authorities, what does it follow from this for the forensic uses of biobanks samples? Suppose we have to deal with a serious crime – one, say, affecting the physical and psychological integrity of one or more persons – and that we find DNA-material at the crime scene, most probably left by the criminal. Suppose further that technical procedures were available (which is not the case at present) that allow to find the perpetrator simply by submitting the sample to existing biobanks. Would it then (ever) be right to look for a sample match in order to identify the offender? Obviously that should depend on the kind of crime as well as on the circumstances in which it was committed. If, for instance, there are other, equally effective, means to identify the offender (for instance, there are reliable eye witnesses or other pieces of evidence), then it does not seem justified to risk public trust on biobanks confidentiality. And when the investigation involves a minor offence we might also not have strong enough reasons to suspend anonymization.[7]

[7] I abstain here from stating where the distinction should be drawn between serious and non-serious criminality, as that is related to complicated value issues. Crimes against persons seem at first sight

But if we had to deal with serious criminality, and no other evidence were available, I believe it is prima facie justifiable to look for a match in available biobanks, even when donors would oppose. The reason for this is rather straightforward. As we have seen, confidentiality is only one aspect of the more complex notion of privacy, in which the protection of individuals' physical and psychological integrity is also included. Obviously, these latter factors are deeply affected by criminal victimization. Particularly the capacity to make autonomous decisions might be disrupted by crime victimization, due to the noxious consequences – both material, physical and psychological – of crime.

So, even if we in some particular situation decided to sacrifice confidentiality to solve a crime, this should not be uncritically seen as a overt violation of privacy, as we would be acting in order to protect other dimensions of the same notion of privacy, namely the physical and psychological integrity of actual and future crime victims.[8]

Thus, according to this approach, instead of adopting an all-or-nothing stance to privacy, we would be required to perform trade-offs between the different aspects included in this notion. This leaves us of course with very little practical guidance.[9] But perhaps a more substantial answer to this query could be afforded if we try to meet the objections to forensic data banks formulated at the beginning of this article. Let us therefore discuss them one by one.

9.6 The Objection from Social Discrimination

Some authors affirm that forensic data banks are socially discriminatory. Thus, for instance, C. Barsby and D.C. Ormerod believe that this practice "arouses fears of the police being able to create (and perpetuate) a category of usual suspects, and for the police to decide who are those free of any taint of suspicion whose samples should be destroyed".[10] This is particularly tenable, as we know that the police does not submit to arrest a representative cross-section of the whole population, but rather chooses its targets among previously identified – or it could even be said

to be a proper candidate to be included in the first of these two groups. However this delimitation should not be seen as conclusive, as other types of criminal conduct usually not conducive to great personal damage might, in particular circumstances, have devastating effects for the victims.

[8]Apparently, this is also acknowledged in the European legislation. Art. 8, 2 of the European Convention on Human Rights states that "There shall be no interference by a public authority with the exercise of this right [to privacy] except such as is in accordance with the law and is necessary in a democratic society in the interests of national security, public safety or (. . .) for the prevention of disorder or crime (. . .) or for the protection of the rights and freedoms of others."

[9]In that sense, it might be said that the right to privacy is a (at least partially) self-annulling right. Unlike other absolute rights that are overridden by other, competing absolute rights, privacy is sometimes overridden by itself: some of its aspects have to be set aside to further another aspects of the same right.

[10]In Barsby and Ormerod (2003, quotation on p. 41).

stigmatized – social groups. Forensic data bases risk therefore to be biased against those regularly arrested. As suspects and criminals often stem from certain social categories (it is after all the poor and unprivileged who most often stand before criminal justice officials), the stored data turn therefore into a register on a particular social group which has no parallel regarding other, better-off groups in society. This discriminatory effect is particularly troublesome when the data gathered are not destroyed once the suspected person is dismissed from the investigation or even acquitted in trial.

To begin with it should be pointed out that the present objection is often directed to the kind of special forensic data bases characterized as (a) before. As the samples are gathered from people already submitted to coercive measures within the legal system, consent procedures and other legal safeguards usually granted to samples donors of medical biobanks are set aside. This however does not apply to the expanded forensic uses of existing medical biobanks (types (b) and (c) above). As donors or their legal representatives voluntarily agree to contribute with a sample, their legal safeguards are not curtailed, at least not to the same degree as the legal guarantees of suspects and offenders are. In that regard, if we intend to reinforce citizens' rights vis á vis criminal justice officials, then the most rational choice is to allow for forensic uses of medical biobanks, instead of creating special forensic data bases. In the former, a court decision will be required to open the register to the police; in the latter, it will often suffice with the decision of the investigating attorney.

But even limited to forensic biobanks of type (a) above, this critique need however to be made more nuanced. As with all other sorts of practice, the objection from discrimination can be easily neutralised simply by enlarging the practice to include the whole population. If everyone is registered, then there is no discrimination. Thus, in that regard, the most promising alternative among the different biobanks mentioned above is gathering biological samples from all new-borns, independently of social and economic status.

Second, we should also keep in mind that forensic data banks aim at solving and preventing serious criminality. Perhaps with the exception of some property crimes (for instance, burglary: it does not seem rational to break in and steal at your poor neighbour's house), most crimes are committed within the same social group.[11] That means that, even if we granted that forensic registers discriminate against a particular social group, this is made with the goal of protecting other people who belong to the same group. Again, we notice that as soon as we scratch beneath the surface of this objection, what we have to confront are difficult trade-offs, rather than conclusive statements.

[11] This point might be illustrated by the US statistics on murder victims. According to a report from the Federal Bureau of Investigation, most murder is intra-racial, not interracial. That means among other things that African-Americans are disproportionately victims of homicide, and that their murderers are also mostly African-Americans. See Cole and Smith (2007).

9.7 The Objection from Miscarriages of Justice

Another objection to forensic uses of biobanks runs as follows: Using genetic material in criminal investigations can lead to wrongful convictions. There are some examples of persons who were wrongfully convicted as a consequence of errors committed in genetic data analysis.

There are however some problems with this objection. In the first place, it is not a principled one. Suppose we could develop sufficiently accurate methods for genetic analysis related to crime scenes. Rates for wrongful convictions might then perhaps be reduced to, say, the same level than that for judicial errors. What could then be objected to forensic data bases? We all seem to accept that some innocent people are wrongfully condemned from time to time. We do of course all we can to minimize the incidence of judicial errors, short of abolishing the criminal justice system, which we deem necessary to avert noxious conduct. In that sense, we believe the (unintended though foreseen) consequence of sending a number of innocent people to prison is still a price worth paying, provided we don't want to risk anarchy and social chaos. I believe something similar might be said on wrongful convictions generated by mistaken genetic analysis. We should of course do whatever we can to avoid them or, at least, to reduce their number. But as long as we deem the goal of achieving high crime clearance rates desirable, we should accept a certain amount of (foreseen though unintended) wrongful convictions.

Second, eyewitness misidentification is the single greatest cause of wrongful convictions nationwide in the USA, involved in more than 75% of convictions overturned by DNA-testing. Does this mean that we should refrain from these testimonies? If the answer is no, why should we then refrain from genetic data analysis?[12]

Third, even if genetic data analysis can be used to wrongly inculpate suspects, they can also exonerate people wrongfully suspected, or even condemned, for crimes they have not committed. As a matter of fact, there is already an important number of cases in which individuals serving long prison terms have been released in the light of new DNA-evidence. According to *The Innocence Project* in USA, the numbers of wrongfully convicted who has been released from prison thanks to DNA analysis is 254, to June 21, 2010.[13] In this regard, the present objection is ideologically biased against forensic data bases, as it only takes up the probable negative consequences of the practice and obviates the already existing positive ones.

---

[12]See *The Innocence Project*, http://www.innocenceproject.org/ (accessed 15 March 2011).

[13]Ibid.

9.8 The Objection from Authorities' Misuse

Another common objection to forensic data banks says that they can be misused by the authorities to control citizens by using previously stored sensitive information.[14] The scenario these objectors have in mind is one in which stored material suggesting psychological or psychiatric disorders can be used as a justification for a penal or a therapeutic measure.

However, at least in its present formulation, this objection is too inclusive. As a matter of fact, something similar could be argued on any sort of register or data gathering by the authorities, either medical or of another kind.

Furthermore, this objection seems to focus on the wrong kind of facts. If someone misuses something, then the problem lies in those who are responsible for the misuse and, ultimately, in the lack of appropriate control mechanisms to prevent it. It is at least not obvious that the problem depends on the very thing being misused, unless we can point out some aspect of this thing that makes it particularly adapted for misuse.

But, it could be retorted, biobanks are different! Unlike standard medical and biological information, genetic data might be used to predict individuals' future conduct on the basis of some observed genetic predispositions. The threat of behavioural genetics – these critics underline – should therefore not be underestimated. This is at least how Tania Simoncelli and Helen Wallace (2010) argue in their article "Expanding Databases, Declining Liberties":

> Expanding these databases puts increasing numbers of people on a "list of suspects" even though they may never have been charged or convicted of a crime. This may subtly alter the way they are viewed both by the state and by their fellow citizens, potentially undermining the principles of "innocent until proven guilty" and of rehabilitation.
>
> Without adequate protections, these permanent records of arrest could be used in future to restrict people's rights and freedoms, for example to make it difficult for them to obtain travel visas or employment.[15]

Now, it is still not clear how biobanks information might be misused, in a way that would be different from other registers. The reference to using these records "to restrict people's rights and freedoms", such as denying them "travel visas or employment", obviously does not hit the mark. This applies as well to

---

[14]The problematic aspects of the storage of biological material from suspects were actualized in 2008 by S. and Marper v The United Kingdom. The case involved two claimants from Sheffield, England: Mr. S. and Michael Marper. Mr. S's fingerprints and DNA samples were taken when he was arrested at the age of eleven and charged with attempted robbery, but he was later acquitted. Michael Marper was instead charged with harassment of his partner. As both suspects became reconciled, they claimed that their previously taken samples were destroyed. The case was brought before the European Court of Human Rights (ECHR), whose decision overturned previous judgements favourable to the government from the United Kingdom's House of Lords, Court of Appeal and High Court. The ECHR held that holding DNA samples of individuals arrested but who are later acquitted or have the charges against them dropped is a violation of the right to privacy under the European Convention on Human Rights.

[15]In Simoncelli and Wallace (2006).

other kinds of registers in which sensitive data are gathered. Employment history, school background, contacts with social authorities, drug abuse records, etc., are regularly used by employers or the authorities to make decisions affecting the persons concerned. Usually we oppose to this practice; in some countries or states, such practices are even banned in the law.[16] If we think it appropriate, we could do the same regarding genetic biobanks information or, at least, submit its use to strict public control. If objectors do not believe in this possibility, they would then have to explain what is so special with behavioural genetics that precludes this way of tackling the problem.

So, to conclude, if the objection is that there is something particularly sensitive in gathering biological or genetic information, critics still have to substantiate this claim by pointing out how State authorities might misuse this information, beyond the standard misuse they could make of other data. So far, however, they have not delivered such argument.

9.9 The Objection from Donors' Drop Out

Finally there is a strong argument against opening clinical biobanks to forensic uses: potential donors' willingness to donate samples might decrease as a result of people's concern about privacy. There is in fact some empirical evidence supporting this claim. In Sweden samples from the Swedish PKU-biobank have been used in a couple of cases for searching suspects in criminal investigations by DNA comparisons.[17] The most discussed case was the investigation after the murder of the former Swedish foreign minister Anna Lind in 2003. The Swedish prosecution authority requested a sample from the suspected murderer from the PKU-biobank. This sample was sent to the Department of Forensic Genetics for DNA-profiling to be compared with traces found at the crime scene. This use of PKU-biobank's material for crime investigative purposes originated a nation-wide debate among experts and the media. As a result of it, about 2000 people requested the destruction of their previously donated PKU-samples.

The Swedish data are indeed worrisome. This possibility of obtaining positive clinical and research results from biobanking depends on the public's willingness to donate samples. Some of the arguments advanced in this article suggest that there might be some reasons to be flexible regarding the handing-over of samples to the prosecution and police investigators. It was argued for instance that such a practice might even strengthen privacy (or at least some aspects of it) rather than jeopardizing it. But in the end the crucial issue is how the public experiences this

[16]For instance, according to the Punishment Sentencing Commission of the state of Minnesota, racial factors are banned in sentencing repeat offenders.

[17]Another example of forensic use of clinical biobanks in Sweden was the identification of deceased Swedish citizens after the Tsunami 2004. For that purpose, a temporary law was passed stating that samples from the PKU-biobank could be used for the identification process. This use of biobanks material did not cause however resistance among the public.

practice. As long as public reactions are as they appear to be, there are strong reasons to keep crime-related and clinical biobanks apart, if we do not want to risk loosing a considerable number of samples.

9.10 Conclusions

The conclusions of this article can be summarized as follows:

- Even if the right to privacy is widely recognized, the most reasonable interpretation says that, although donors have a legitimate claim not to get sensitive information about them disclosed without their permission, this right might in some circumstances be overridden by competing claims of more social, legal or moral weight. Accordingly, rather than simply referring to the right to privacy as an objection to forensic biobanks, it would be much more fruitful to ask: *Which kind of right is the right to privacy?* My answer to that question was that it is an absolute though overridable right.

  Particularly in criminal contexts, donors' right to privacy have to be weighed against a corresponding right to privacy of both actual and potential crime victims. To put it shortly: we would have to choose between either sacrificing confidentiality (which mainly regards biobanks' donors) and individuals' physical and psychological integrity (particularly related to crime victims). Both these aspects are – it should be recalled – included in the general notion of privacy. Accordingly, rather than asking whether privacy can be legitimately violated for crime investigative uses, the issue to be raised is: *Whose privacy should be protected, and in which circumstances?*[18]

  With these distinctions in mind, the traditional objections to forensic uses of biobanks were met as follows:
- Regarding the objection from social discrimination, the solution is implementing a large scale population genetic data bank. If all citizens are in the register, then there is no discrimination. Furthermore, even regarding targeted population data bases, they can also be seen as tools implemented to protect the same social group that is targeted, given the fact that most crimes are internal to the group.
- Regarding the objection from wrongful convictions, I argued that this argument rather speaks in favour of genetic data banks: through developments in genetic analysis techniques, the error rates could be reduced, innocents acquitted and legal security therefore increased.
- Finally, concerning the criticism on misusing biobanks samples for government surveillance, I argued that objectors still owe us an account of how this misuse would differ from a similar misuse of other widely accepted registers. They also

[18]This formulation is obviously not totally accurate, as some donors will certainly also become crime victims in the future. For reasons of simplicity, however, I choose to keep this formulation.

owe us an explanation of why they believe this risk cannot be averted through the implementation of appropriate mechanisms of public control.

The upshot of this discussion is that the question of whether or not it might be justified to (ever) violate donors' right to privacy in biobanking practices for crime investigative purposes is much more difficult to answer than what is usually presupposed in the legal and ethical debate. As a matter of fact, the question is not even correctly formulated, as the conflict is not between privacy and some other valuable social goal, but instead between different aspects of the very notion of privacy. Or so have I argued. The most reasonable stance is probably to see donors' right to privacy as absolute while leaving the decision whether or not to override this presumption in a particular criminal case to a court of law. Further weakening this right, for instance making it just a prima facie one, might negatively affect people's willingness to become donors.

These arguments however have to be weighed against existing empirical evidence (mainly the Swedish experience) suggesting that potential donors might become reluctant to contribute or actual donors prone to withdraw, once they come to know that crime investigative uses are not excluded as a matter of principle, but could be made possible by a court decision. This issue is of vital importance, as the possibility of materializing the benefits of biobanking depends upon a large sample collection. Thus, provided the public's reactions to using clinical biobanks samples for crime investigative purposes remain as they apparently are, there are strong reasons for keeping these two types of biobanks apart.

References

Barsby, C., and D. C. Ormerod January 2003. "Case Comment: Evidence: Retention by Police of Finger Print and DNA Samples of Persons Subject to a Criminal Investigation but not Subsequently Convicted." *Crim. LR*: 39–41.

Beckman, L. 2005. "Democracy and Genetic Privacy: The Value of Bodily Integrity." *Medicine, Health Care and Philosophy* 8: 97–103.

Case of S. and Marper v. December 2008. The United Kingdom. "ECHR." Accessed March 3, 2011. http://en.wikipedia.org/wiki/S_and_Marper_v_United_Kingdom#Background().

CDBI. 2006. *Draft Explanatory Memorandum to the Draft Recommendation on Research on Biological Materials of Human Origin*. Strasbourg: Council of Europe Steering Committee on Bioethics.

Cole, G. F., and Smith, C. E. 2007. *The American System of Criminal Justice*, 11th edn. Canada: Wadsworth Cengage Learning (particularly Chapter 2, "Victimization and Criminal Behavior", pp. 41–42).

Council of Europe. 2006. *Recommendation Rec (2006)4 of the Committee of Ministers to Member States on Research on Biological Materials of Human Origin*. Strasbourg: Council of Europe.

DeCew, J. W. 1999. "Alternatives for Protecting Privacy While Respecting Patient Care and Public Health Needs." *Ethics and Information Technology* 1: 249–55.

Elger, B., and A. Caplan 2006. "Consent and Anonymization in Research Involving Biobanks." *European Molecular Biology Organization Reports* 7 (7).

Simoncelli, T., and H. Wallace 2006. "Expanding Databases, Declining Liberties." *Gene Watch* 19 (1), January–February. "Council for Responsible Genetics." Accessed March 3, 2011. http://www.councilforresponsiblegenetics.org/GeneWatch/GeneWatchBrowser.aspx?archive=yes&volumeId=19

Teetzel, S. 2009. "Respecting Privacy in Detecting Illegitimate Enhancements in Athletes". In *The Ethics of Sports Medicine*, edited by C. Tamburrini and T. Tännsjö, 41–52. London, New York: Routledge.

Von Hannover v Germany. Case reference [2004]. EMLR 379; (2005) 40 EHRR 1 [2004] EMLR 379; (2005) 40 EHRR1.

Part III
Regulation of Tissue Research

Chapter 10
A Unified European Approach on Tissue Research and Biobanking? A Comparison

Katharina Beier and Christian Lenk

10.1 Introduction

Starting from the observation that the regulative heterogeneity of human tissue and biobank research is virtually common sense in the literature (see Boggio et al. 2007; Kaye 2006; Maschke 2005; Cambon-Thomsen et al. 2007), this article aims to step beyond this rather descriptive approach. In particular, we will not only highlight common trends and perspectives in the regulation of human tissue research across the countries of the European Union and Switzerland but we will also identify the ethical and legal foundations for some of the persisting differences in this field.[1] Thereby our analysis bears on the premise that certain countries hold similar research traditions and are also united by common ethical and legal pathways for regulating research. However, this is not to neglect the prevailing differences in this field, but rather to avoid their exaggeration. For this purpose, our argumentation will proceed as follows: Firstly, we will provide an overview on the most significant European documents pertaining to human tissue research and biobanking (II). Secondly, it is the aim of the article to outline common trends in the national regulations by looking at country groups that approach issues of human tissue research in a similar way (III). On the basis of these findings we will finally draw some conclusions regarding the future regulation and potential legal harmonization of this field within the European Union (IV).

[1]This article is based on findings of the EU-funded Tiss.EU project "Evaluation of Legislation and Related Guidelines in the Procurement, Storage and Transfer of Human Tissues and Cells in the European Union – an Evidence-Based Impact Analysis" that is coordinated by the Department for Ethics and History of Medicine at the University of Goettingen, Germany (www.tisseu.org).

K. Beier (✉)
Department of Ethics and History of Medicine, University of Göttingen, Göttingen, Germany
e-mail: katharina.beier@medizin.uni-goettingen.de

C. Lenk et al. (eds.), *Biobanks and Tissue Research*, The International Library of Ethics, Law and Technology 8, DOI 10.1007/978-94-007-1673-5_10,

10.2 State of the Art: European Documents for the Regulation of Human Tissue Research and Biobanking

In the European context the following documents are of major importance in the field of biobanking and human tissue research: The *Convention on Human Rights and Biomedicine* of the CoE which was signed in Oviedo in 1997.[2] The Convention, amongst others, requires express and informed consent for the accomplishment of research (art. 5 and 16).[3] Furthermore, the misuse of genetic data for the discrimination of persons (art. 11) and financial gain resulting from the usage of human body parts are prohibited (art. 21). Of particular importance in the field of tissue research is the provision that a "secondary use" of body material is only acceptable with the patient's or proband's informed consent (art. 22).[4] In addition (in accordance with the EU Data Protection Directive), every person has the right to know the information which are collected and stored about his or her health condition (art. 10).

The European Data Protection Directive (1995/46/EC) is pertinent to the field of human tissue and biobank research on the account that it governs the processing of all personal data. However, it is at least questionable whether the Directive actually applies to all data obtained in the course of biomedical research as it can be argued that these data do not equally present a threat to the source's privacy.

The first official European document on *research* with human biological materials[5] stems from the Committee of Ministers of the CoE which released its *Recommendation 2006 (4) on research on biological materials of human origin* in 2006. In particular, the Recommendation contains provisions on the identifiability of samples and data (art. 3), the use of residual materials (art. 12), the issue of population biobanks as well as the use of biological materials in research projects (art. 21 sf.). Furthermore, the Recommendation gives some orientation regarding organisational issues, for example the responsibility for the collection (art. 14 sec. 1); rules for the collection's purpose and the access, use and transfer of samples (art. 14, secs. 2 and 4), requirements for documentation of the samples' origin and donors' consent (art. 14, sec. 3), rules for the establishment of population biobanks (art. 19, secs. 1 and 2) as well as independent oversight on research purposes (art. 24, sec. 1).

However, although the CoE's Recommendation was meant to offer legal guidance to the European Member States, it did not succeed in abandoning the legal insecurity. For example, there are several overlaps with the *Additional Protocol to the Biomedicine Convention of 25 January 2005 on research with human beings*

---

[2] Whilst the Convention is not ratified by all Member States of the European Union so far, the CoE has adopted it.

[3] See also the UNESCO International Declaration on Human Genetic Data (2003, art. 8a).

[4] See also the Convention on Human Rights and Biomedicine (1997, art. 22).

[5] It is important to note that the Tissue Directive 2004/23/EC merely covers therapeutic applications of human biological samples.

(e.g. with regard to informed consent for the removal of human biological material for research storage) (see Nys 2008). Given that the Recommendation provides only for basic rules which leave considerable leeway for its transposition into national law, the regulative "state of the art" regarding human tissue and biobank research becomes only understandable through an in-depth investigation of national laws and guidelines in this field (see Boggio et al. 2007). In comparison to the latter's, however, our work is different in two aspects: Firstly, whilst Boggio et al. consider biobank guidelines and regulations world-wide, our focus will be exclusively on European Member States, including Switzerland. Secondly, whereas Boggio et al. have chosen a "topical" approach by citing the regulations and guidelines of certain countries as examples of different positions on a range of human tissue research issues, we will base our investigation rather on a genuine "regional" approach; i.e. by laying the focus on country groups with resembling research and legal traditions, we will analyse their respective way of coping with four decisive ethical and legal challenges of human tissue and biobank research.

10.3 Common Trends and Perspectives in the Regulation of Human Tissue and Biobank Research

10.3.1 Central Issues of Human Tissue and Biobank Research

Although there is a wider array of concerns that can be raised in the context of human tissue and biobank research, we will concentrate on four issues on the account that they are also addressed by many national frameworks and guidelines dealing with human tissue and biobank research.

(a) *The issue of property and ownership in human tissues and cells* is intensively discussed since the establishment of large-scale tissue collections challenges the widely acknowledged position that the human body and its parts are "*res extra commercium*".[6] In particular, given that human tissues are increasingly collected, stored and processed for research purposes and may generate material as well as immaterial benefits to researchers and/or research conducting companies (for example pharmaceutical enterprises), the legal status of separated bodily materials calls for clarification.
(b) The obtainment of *informed consent* from potential research participants is the default position across the European Member States. However, due to the peculiar features of human tissue and biobank research the situation is getting more complicated. Challenges arise from two sides. Firstly, the meaning of the word "informed" can be called into question. The prospective character of biobank research makes it difficult to inform the participants at the time of donation of their bodily materials exactly on the potential uses and risks connected with

[6]See also the Oviedo Convention or the CoE's Recommendation 2006(4), art. 21.

this kind of research. As a matter of fact, newly emerging research questions might make it desirable to use samples for purposes lying beyond the initially obtained consent (secondary use). Secondly, these challenges are often faced by taking refuge to an expanded understanding of "consent" defined as *broad, open, blanket* or *general consent*, or by replacing the donor's consent by the concept of *presumed consent*. The CoE's Recommendation does not provide for a clarification of this aspect as it only states that "information and consent should be as specific as possible" (art. 10). In our analysis we will therefore ask whether the respective national legislations stick to the traditional rule of informed consent or whether they allow for exemptions.

(c) The protection of the sample donors' *privacy and confidentiality* in the field of human tissue research is a particularly sensitive issue due to the fact that besides samples also personal data are collected, stored and processed. Although anonymisation may help to protect the source's rights to privacy and confidentiality, researchers often perceive this as a too cumbersome method. Given that many research projects depend on the traceability of samples or the identifiability of sample donors respectively, pseudonymisation, i.e. the coding of biomaterial and health data, is regarded more suitable from the researchers' point of view. However, the provisions which kind of samples and data and under what conditions need to be anonymised or pseudonymised vary considerably across the European Member States (see Sándor and Bárd, Chapter 14, in this volume).

(d) The issues of *benefit sharing and feedback of health-related information as a particular means of benefit sharing* arise due to the fact that human tissue is not only valuable for therapy but also a source of economic gain. This poses the question whether it is justifiable that the participants of human tissue research provide their bodily material for free. Another aspect that is closely connected with the issue of benefit sharing is the question whether there is an obligation of researchers to give feedback on health-relevant findings to tissue donors. Whilst from an ethical point of view it is problematic to retain health information with therapeutic relevance (Johnston and Kaye 2004; Lenk 2011), only few jurisdictions tackle this issue head-on. The *European Data Protection Directive* stipulates a person's right to gain information on her stored data (art. 12). Even though it does not mention medical or health-related data explicitly, it is quite reasonable that this kind of information falls under the scope of this Directive, too.

10.3.2 Identification of Country Groups and Regulative Approaches

Coming to the identification of European country groups, we will base our analysis on a regional-political grouping. The rationale behind this structure lies in the assumption that there are common ethical and legal traditions in the respective country groups.[7] Against this background, we will firstly deal with, the German speaking countries, as there are Austria, Germany, and Switzerland (as the only non-EU state included here); secondly, the Benelux states, consisting of The Netherlands, Belgium, and Luxembourg; thirdly the Anglo-Saxon countries, comprising the UK and the Republic of Ireland; fourthly, Sweden, Denmark, and Finland which represent the so-called Scandinavian countries. The fifth group comprises the three Baltic States, i.e. Estonia, Lithuania and Latvia, whereas sixthly the countries bordering the Mediterranean Sea include France, Spain, Portugal, Italy, Cyprus, Malta, and Greece. Finally we will deal with the group of Eastern European States that entered the European Union in 2004, or 2007 respectively. This group consists of Slovenia, Slovakia, Czech Republic, Hungary, Poland, Romania, and Bulgaria.

As a last prerequisite for the following analysis we distinguish between four regulative approaches: Firstly, there are virtually some "pioneer" countries which regulate human tissue and biobank research by a discrete law. In particular, these are Sweden, the UK, Portugal, Spain, Estonia, Hungary, Belgium, Lithuania and Latvia. From this group we can secondly distinguish those countries that are just about enacting laws, as there are Switzerland, Finland, Slovenia, and Poland. The third group of countries tackles the field of human tissue and biobank research by a rather fragmented jurisdiction, entailing references to the Civil and/or Penal Code, Constitutional law, Codes of professional conduct, transplantation-, transfusion-, funeral-, drug- as well as data protection laws, whilst there are no clues for this fragmentation to be abandoned in the near future. However, with the exception of Luxembourg and the Netherlands, in the countries belonging to this group discussions on human tissue research and biobanking have taken place and official opinions of (Bio-)Ethics Committees or similar bodies are available. In particular, this is the situation in Germany, Austria, Greece, Cyprus, Italy, the Republic of Ireland, France, and Denmark. Finally, there is the group of Eastern European Countries and the more recent accession states respectively, including Romania, the Czech Republic, Malta, Bulgaria, and Slovakia, that leave the field of human tissue and biobank research rather unregulated. In contrast to the third group, a public debate was hardly held and no official opinion of a Bioethics Committee has been laid out so far; rather, legal guidance in this field is mainly derived from transnational legislation and codes, like EU Directives (e.g. the Tissue Directive) or the Oviedo Convention.

10.3.2.1 The German-Speaking Countries

In the German-speaking countries human tissue research is not regulated by a distinct law so far, but subject to provisions of transplantation-, transfusion-, drug- and funeral law. Further clues to the regulation of this field can be derived from

[7] This classification resembles the structure of the Tiss.EU project. For more information on its research methodology, see www.tisseu.org (accessed 4 March 2011).

opinions of the German Ethics Council (NER 2004/DER 2010)[8] and the Bioethics Commission at the Austrian Chancellery (2007).[9] In Switzerland the recommendations of the Swiss Academy of the Medical Sciences (SAMS 2006)[10] as well as the proposition for a comprehensive Human Research Act (HRA) are instructive in this matter.

Regarding the legal status of human bodily materials, the three countries take an obviously similar stance. Although tissues and the living human body in general are perceived as a matter beyond property, the "source" of the tissue can still claim a de facto property-right as soon the material has been separated from her body. Justification to this position is derived from the donor's personality rights. According to this approach, extracted human materials are considered as "continued personality". From this it follows that donors do not only have the material disposal of their tissue sample, but – as it is part of their person – may retain further rights, for example, the right to decide on their tissue's utilisation for therapy or research. Having said this, Germany, Switzerland and Austria do not allow for an automatic transfer of property in human tissues, but donors can only transfer their rights by explicit consent. The situation, however, is far less clear, if it comes to remaining bodily materials as leftovers from surgeries or biopsies. Previously these materials were often employed by the hospital without the patient's consent or by simply presuming her "silent consent." In light of large systematic repositories that can be used for genetic analysis, however, these practices appear doubtful from an ethical point of view.

In Germany for research with identifiable samples, the donor's informed consent displays the default condition. However, the Ethics Council pleads for exemptions from this rule in the case of anonymised samples and argues that also a broad consent should be acceptable (NER 2004, 14/DER 2010, 4.2.2). In Switzerland according to the SAMW guidelines (2006, 4.3) and the proposition for the HRA, donors may even give their general consent (*Generalkonsent*) for the application of their samples and data in future research projects. In this regard, the Swiss approach is perceived as utmost research-friendly (see Dörr 2011). While the Austrian Bioethics Commission (2007, VII., 2.1.1, 76) recommends the donor's informed consent, which is only valid on the condition that it comprises the obtainment, application, *and* processing of samples and data, in contrast to its neighbour countries, no general consent is foreseen.

As the German-speaking countries emphasize the protection of the donor's personality rights, it is only stringent that the forthcoming Swiss HRA makes the feedback of health-relevant information obligatory, but at the same time acknowledges a person's right of *not* being informed (2006, art. 11).[11] The right of feedback

[8]Biobanks for research (2004); Human tissue biobanks for research (2010).

[9]Biobanks for medical research (2007).

[10]Biobanks: obtainment, preservation and use of human biological material. Medico-ethical guidelines and recommendations (2006).

[11]This is also required by the guidelines of the SAMS (see 2006, 4.7).

on health information is similarly set forth by the German Ethics Council (2004, chap. 5). The position in Austria is slightly different: according to the Bioethics Commission there is no individual duty of information in general, but only in the case that the information is essential for the donor's life (2007, VII, 3.4). However, as this wording is not explained any further, a broad understanding of "essential for the donor's life" might still allow for the provision of feedback on health findings.

As a further common denominator, the protection of the donors' privacy is emphasized by all three countries. To this end, the German Ethics Council proposed the introduction of a biobanking secrecy (DER 2010, 4.2.1). Furthermore, like the Austrian Bioethics Commission and the Swiss SAMS, it prefers the coding of all personal features that allow for the identification of the donor (as a means of reversible anonymisation, i.e. pseudonymisation)[12] to the means of irreversible anonymisation. The Swiss proposition for the HRA even prohibits the anonymisation of samples provided that the results of the research project can be expected to prevent or cure diseases of the respective participants (see HRA, art. 14).

In a nutshell: the German-speaking countries' regulative approaches coincide on most aspects. Although Switzerland is about enacting a specific human research law, the proposition for the HRA does not indicate any serious deviation from the regulative principles applied in Germany and Austria so far. However, given the German Ethic Council's recent recommendation for a specific regulation of human tissue biobanks for research (DER 2010), sooner or later the German-speaking countries might decide to abandon their fragmented approach in this field.

10.3.2.2 The Anglo-Saxon Countries

Due to diverse disclosures organ retention scandals in the United Kingdom and Ireland, these countries reworked and actualized not only their handling of removed organs, but also of human tissues for research fundamentally. In the organ scandal at Liverpool's Alder Hey Hospital, doctors were for example accused of lying to parents regarding autopsy procedures and the retaining of organs from their children (Anonymous 2002). In the Irish *Madden Report* (Madden 2006), it was revealed that "it was not hospital or professional policy to inform parents that in the course of a post mortem to be carried out on their child, organs may be retained, stored, and subsequently disposed of." As a consequence, issues of information of patients and relatives regarding the removal and transparency regarding the use of organs and tissue in general became far more important than in the past. As José Miola (Miola 2011, chap. 8.3) points out, the former legislation in the UK (i.e., the Human Tissue Act 1961) was largely unsatisfying due to two reasons: first, it failed to clearly define who has the right to authorize the removal and retaining of organs due to a post mortem. Second, although some conditions for the lawful removal of organs and tissue have been laid out, there were no sanctions defined in case of a breach of

[12]As an additional precaution it is required that the key-code is kept separately from the coded data.

these prescriptions. For this reason the donors' autonomy and the accountability of medical professionals were the two decisive guiding principles for the establishment of the *Human Tissue Act* (HTA) in 2004 (see Miola 2011, chap. 8.4). According to David Price, with the introduction of the HTA in the UK a shift from a de-facto opt-out-system to an official opt-in-system has taken place that will be irreversible for the nearer future due to the experiences from the organ retention scandal (see Tiss.EU Report Birmingham 2010).

In its opinion *Human biological material: Recommendations for collection, use and storage in research* (2005), the Irish Council for Bioethics requires that "research involving children should only be carried out when the research cannot be equally well carried out with adults and its purpose is to obtain knowledge relevant to the health needs of children" (Rec. 5). Additionally, it is recommended that research with minors should be "of negligible risk and not unduly invasive" (ibid.). Given that for archived and anonymous biomaterial it is often not possible (and therefore also not necessary) to obtain the donor's informed consent, Rec. 10 states that researchers should then seek the approval of the local REC. In contrast, in the case of identifiable material and personal data (this means also coded or pseudonymised tissue samples or patient records), research should be done only with the donor's information and consent (Rec. 11). The status of a tissue sample which is transferred from the donor to a research institution is named a "gift" in Rec. 12. This notion is obviously used to justify the denial of any financial remuneration to the donor (which is in accordance with the respective European guidelines). The requirements for confidentiality and data protection of the Irish Council for Bioethics (see Rec. 17 and 18), are in line with the EU Data Protection Directive. Rec. 20 also demands the information of the donor or study participant in case of research findings of "immediate clinical relevance to a participant".

This is in clear contrast to the situation in the UK as well as to the provision of the *UK Biobank Ethics and Governance Framework* (2006, I B, 3), which expressly excludes feedback in the case of genetic findings which stem from the project's scientific work. By building on the donor's trust, the UK Biobank further provides for a broad consent rule (ibid., I, B, 1). Whilst this can be seen as a considerable movement away from informed consent, this is also true for the HTA. Although initially a specific consent was foreseen, due to pressures of the research community, finally a general consent system has been adopted (see Jose Miola's presentation in the Tiss.EU Report Hannover 2008).

It is an interesting feature of the Anglo-Saxon *Common Law* tradition that the human body (like in the Continental *Civil Law* tradition) is seen as *res extra commercium*. For example, it was argued in the first instance of the British *Yearworth Case*[13] that six men, whose sperm was destroyed due to inappropriate storage, did not have ownership in their (own) sperm, what was only logical according to this original approach. Despite the generally more pragmatic *Common Law* approach, it was up to now even stricter than the *Civil Law* tradition. According to the latter,

[13] *Jonathan Yearworth & ors v Bristol NHS Trust [2009] EWCA Civ 37* (4 February 2009).

in many European countries samples or body parts which are extracted from the human body become things in the legal sense and can be owned and sold to other parties.

10.3.2.3 The Benelux States

The Benelux states (Belgium, Luxembourg, and the Netherlands) display a very homogeneous country group with a similar history and comparable traditions. For example, in the debate on euthanasia and assisted suicide, Belgium and the Netherlands definitely share a similar approach. However, in the field of biobanking and tissue research it seems to be difficult to identify common tendencies across these countries. The Netherlands and Luxembourg feature a regulation which resembles rather the German than the Belgian situation. The relevant prescriptions and guidelines are scattered about a number of laws and law sectors. For example in Luxembourg, guidance can be derived from the acts on autopsy and cadavers (1958), on blood (1979), on the removal of human substances (1982) and on tissue and cells (2007).

From the Benelux countries, only Belgium has a discrete human tissue research and biobanking law, the *Law of 19 December 2008* regulating the procurement and the use of human bodily material for medical application in humans or for scientific research. Given that this law displays the implementation of the EU Tissue Directive, Belgium is the only European country which extends the Directive's scope to the area of research with tissue and cells.

Regarding property and ownership in human tissue and cells, in The Netherlands there is usually no formal transfer of ownership from the patient (the donor) to the biobank or research institution, but only a statutory permission of the patient to use the tissue sample for medical research. In contrast, in Belgium biological materials become a "good" after the separation from the human body which is then owned by the biobank. A unique Belgian feature is the *Royal Decree* 14 October 2009, which fixes prices of human material. This approach can be interpreted as a middle way between a strict non-commercialisation of body material and the acceptance of market prices for human tissue. The price for human tissue thereby underlies the supervision of state authorities.

With the establishment of an opt-out rule in the case of deceased donors there is also a special path taken in Belgium. In line with the *Organ Donation Law* from 1986, tissue and cells can be taken from the deceased person when this was not excluded during her life. However, in the case of using tissue from living donors, for research purposes the informed and written consent of the patient or proband is legally required. The secondary use of tissue samples is shaped in a more liberal way. The patient is informed in writing of the possibility of secondary use. Provided that he does not object to this research, his consent is assumed (art. 20, §2). In The Netherlands, a distinction is made between anonymised tissue and tissue samples which are linked to the donor: while in the former case it is sufficient to inform the patient about the scientific use of his bodily material, and he does not object to this use, in the latter case one has to obtain his explicit consent. In contrast to The Netherlands and Luxembourg, in Belgium the donor is granted an explicit "right to be informed" of relevant health findings. Art. 11 of the Belgian Law further stipulates that "the physicians who learn such information in the context of an action with or use of the material, the governors of the human body material and the chief physician of the hospital where the removal took place are, each in the context of their function and role, responsible for the application of [this provision]."[14]

It can be concluded that many important issues are left unregulated in The Netherlands and Luxembourg, e.g. feedback of health relevant information (only the Belgian Law foresees a right of information for the donor), ownership, benefit sharing, etc. Thus, at the present point of time, much insecurity remains – not only for patients and tissue donors, but also for researchers and their institutions.

10.3.2.4 The Scandinavian Countries

The Scandinavian countries represent an exceedingly homogeneous country group with shared traditions and values, e.g. the reliance on public deliberation or the Scandinavian model of welfarism. Regarding human tissue research, there is a long-standing tradition in health registries and epidemiological research, which is facilitated by the existence of unique personal identification numbers. Consequently, Scandinavia is known as a downright research-friendly area.[15]

Despite these common traditions, there is surprisingly little convergence in the regulation of human tissue and biobank research in Sweden, Denmark, and Finland (see Rynning 2009). Whilst Sweden enacted a discrete biobank law as early as 2002, Denmark merely introduced complementary provisions into already existing laws[16]. Finland adopted an awaiting stance; i.e. whilst up to now human tissue and biobank research was subject to a rather fragmented jurisdiction,[17] a specific law is expected to be issued in the near future.[18] Beyond these different legal pathways,

[14] Cited after the presentation and translation from Sigrid Sterckx (see Tiss.EU Report Birmingham 2010).

[15] According to figures from the Public Population Project in Genomics (P3G), out of 82 national research cohort studies (comprising 7.8 million subjects), 2 million come from the Scandinavian countries. This matches 25% of the subjects currently enrolled worldwide (see Nobel 2008, 13).

[16] See the report of a Ministry task group (Betaenkning no. 1414, May 2002). Amendments have been carried out on the *Act on the Legal Status of Patients* (482/1998), the *Act on Processing Personal Data* (429/2000) and the *Act on Ethics Review of Scientific Research* (69/1999). The new regulations on biobanks were published as Ministerial Order No. 966 of 22 September 2004 (*Use of Tissue Register*) by the Ministry of Health.

[17] See for example the Finish *Act on the Status and Rights of Patients* (785/1992) and the *Act on Medical Research* (488/1999). However, like Denmark also Finland revised its existing legislation, e.g. the *Act on the use of Human Organs and Tissues for Medical Purposes* was supplemented with provisions on the collection and usage of human tissues (see Rynning 2009, 297).

[18] A proposition for a Tissue Act has been submitted to the Parliament in spring 2010 and is expected to come into force in 2011.

however, Sweden, Denmark, and Finland add up to the same practical result, i.e. the enforcement of an outright research friendly framework.

In particular, although the Swedish *Biobanks in Medical Care Act* (BMCA) is said to provide for one of the strictest consent rules in Europe or even the world (see Dillner 2002; Hansson and Björkman 2006, 285), this is obviously not detrimental to research (see Beier 2009, 2011). For example, the *Personal Data Act* (1998, 204) allows for exemptions from the BMCA's consent rule. Regarding statistics and research purposes, section 19(1) requires the weighing of individual against collective interests: if the latter are sufficiently strong and individual risk is estimated to be low, consent may be waived.[19] This provision does also apply to the obtainment of secondary consent in case of new research purposes, provided that a Research Ethics Committee approves the respective project (see BMCA, chap. 3 (5)). In Sweden, however, the question who is allowed to access the samples and data stored in a biobank attained particular attention when the police accessed the newborn-screening biobank for the conviction of the murderer of Anna Lindh, the former Swedish foreign minister.

Unlike Sweden, Denmark relies on an opt-out system. According to the *Act on the Processing Personal Data* (429/2000) that has been validated to human tissue research, no explicit consent of the subject source is required, but researchers are free to use stored samples unless the donor is registered in an opt-out registry (see Ursin et al. 2008, 181). Similar to Sweden, in Denmark the Research Ethics Committees have far-ranging competencies regarding the definition of informed consent rules. In practice, their allowance for exemptions appears rather as the normal instead of the exceptional case (see Dillner 1999), which makes Denmark a favourable model even to Swedish researchers (see Nobel 2008, 9).

Although in Finland all records and personal data incurred in the context of biobanking are perceived as sensitive data, regarding the surrender of personal data for research, the *Personal Data Act* (523/1999) and the *National Health Registries Act* (556/1989) take a principally liberal stance. This is also true for the obtainment of informed consent. According to the *Personal Data Act* (§14, 1), the processing of personal data for research purposes does not require the consent of the data subject if this "cannot be obtained owing to the quantity of the data, their age or another comparable reason" (research exemption). In addition, the envisaged Biobank Act is expected to switch to a broader consent rule regarding the use of future collections (see Tiss.EU Country Report, Stockholm 2010). A relaxation of consent requirements is also expected after the amendment of the Swedish BMCA.[20] However, given the wide public support of human tissue and biobank research in Scandinavia,

[19]This is a unique feature of the Swedish Data Protection Law which derives support from the EC Data Protection Directive's provision that "member states may, for reasons of substantial public interest, lay down exemptions" regarding the protection of personal data (art. 8, para. 4).

[20]In Sweden a revision of the BMCA is ongoing. In November 2010 the Swedish government released a report "A new Biobanks Act" (SOU 2010: 81) that formulates proposals for an amendment of the present law. For example, it suggests the introduction of an opt-out system for the procurement and storage of samples.

it might be of minor importance whether a strict consent approach, like in Sweden, or an opt-out model, like in Denmark, is applied.

10.3.2.5 The Baltic States

The three Baltic States are known for their innovations in the field of information- and biotechnologies in the post-communist era. In particular, Estonia is not only the inventor of Skype and E-voting in public elections, but like Latvia and Lithuania it has also started a population-wide biobank project. To regulate the biobank's establishment and operation, Estonia and Latvia have adopted specific legislations – the *Human Genes Research Act* (HGRA) and the *Human Genome Research Law* (HGRL) respectively; in Lithuania human tissue research is subject to a more comprehensive *Law on Ethics of Biomedical Research* (LEBR). It is striking that the countries' national biobanks are solicited as a competitive economic and scientific location factor (see Eensaar 2008, 56). Consequently, Estonia excludes any transfer abroad, but samples have to "be stored in the territory of the republic of Estonia" (HGRA 2001, §18 (4)). As a further common denominator, the Baltic legislations build on a relationship of trust with donors (see Salter and Jones 2005) by insisting on privacy safeguards in the field of tissue research. Finally, all three countries have regulations on the issue of feedback regarding health-relevant findings.

The Estonian HGRA was particularly designed to govern the establishment of the national biobank which aims to collect samples as well as medical and genetic data from one million Estonian people. Although donors do not receive any direct financial benefit in neither of the Baltic States, in Estonia GP's obtain a €32–34 financial compensation per donor as an incentive (see Eensaar 2008, 65).[21] Regarding the regulation of consent, the Genome Project provides for an open consent; i.e. donors agree that their samples and related data are not only entered into the biobank, but may also be used for "genetic research, public health research and statistical and other purposes in accordance with the law".[22] According to Ants Nomper, this provision has its roots in the application of the European Data Protection Directive to research with human tissue samples. Compared to consent requirements in medical research, this allows for more relaxed provisions, implying a waiver of consent as well as less strict information requirements. In the recently started E-health project informed consent is even abandoned altogether by making participation mandatory (see Tiss.EU Report Budapest 2009). Another Estonian peculiarity concerns the issue of ownership. According to the HGRA (§15, 1), "the chief processor's[23] right of ownership of a tissue sample, description of state of health, other personal data and genealogy is created from the moment the tissue sample or personal data is provided or the moment the state of health or genealogy is prepared." On the other

[21] For further information on the Estonian framework, see also Sándor and Bárd (2009c).

[22] See Gene Donor Consent Form, available at: http://www.geenivaramu.ee/index.php?id=100 (accessed 4 March 2011).

[23] The chief processor of the Estonian Genome Project is the University of Tartu.

hand, donors are not left without rights. For example, they are granted the right to know their genetic data but may also renounce this right (HGRA, §12 (4), 4, 5).

Whilst in Latvia, informed consent displays the default condition for conducting biomedical research, for becoming a gene donor in the national Genome Project, participants give a rather broad consent which allows their tissue samples, health information and genealogy to be used for genetic research (even outside Latvia) as well as public health and statistical purposes (see Sándor and Bárd 2009b, 3). On the other hand, to assure the participants' trust, they are granted a right to access their genetic data. To this purpose, the HGRL (sec. 20(2)) allows for the decoding of samples. For the benefit of the donor in an emergency situation, however, the treating physicians may receive genetic information without the donor's consent.

Like the Latvian framework, the Lithuanian LEBR (lastly amended in 2007) foresees the obtainment of the donor's informed consent for biomedical research (art. 4; 8(1)). However, exemptions for research with tissue and genetic material can be granted by the Lithuanian Bioethics Committee or a Regional Biomedical REC (art. 8(2)). For an informed consent being valid, it is additionally required to provide information about "foreseeable benefits of the biomedical research to the subject" (art. 8(1)). Although individual health-related information are perceived as confidential, the Lithuanian law allows its disclosure "without the subject's consent if the subject's identity remains undisclosed after such information is made public" (art. 9(2)).

Given the long-term character of the Baltic biobanking projects, it does not come as a surprise that coding which allows for de-identification of donors is favoured by the three countries' legal frameworks. However, regarding coding mechanisms strict safeguards apply. In Latvia, for example, after coding, the chief processor needs to transfer the code, which "shall be the only possible decoding key" to the State Population Genome Register (HGRL §19(3)). In Lithuania, the *Law on Legal Protection of Personal Data* (art. 12) allows for the processing of personal data only after the person's consent; otherwise the approval of the State Data Protection Inspectorate is required and personal data must be altered in such a manner that identification of the person is impossible.

Taken together, the Baltic States' legal frameworks facilitate the accomplishment of large-scale population biobanking. On the other hand it is striking that these projects do not simply follow the logic of economy but are also concerned about the donors' trust in research. In particular, similar to the establishment of the UK Biobank, the dialogue with the public has been sought in order to avoid the pitfalls of the Icelandic genetic database.

10.3.2.6 The Countries Bordering the Mediterranean Sea

Whilst in the Mediterranean region Spain and Portugal feature specific laws on human tissue and biobank research, in France, Italy, Cyprus and Greece the most important issues are regulated by a rather fragmented legislation. Clarification on moot issues can also be derived from opinions of the respective Ethic Committees.

In Malta, however, guidance on human tissue research is only given by transnational Codes and European guidelines. Irrespective of these different approaches, also the countries at the Mediterranean Sea display several commonalities in the regulation of human tissue and biobank research.

To start with, France, Spain and Portugal generally share the no-property approach to the human body and its parts. However, like in the German-speaking countries, in Spain and Portugal the donor retains several rights in his extracted tissues. For example, the Spanish *Biomedical Research Act* (BRA 2007, 70.1) grants the source subject the right on the samples' use and accession. Given the perception of the human genome as common patrimony, Portugal extends the rights on samples and associated data even to the source's relatives. Accordingly, the Portuguese *Law Nr. 12/2005 on Personal Genetic Information and Health Information* (PGIHI) sets forth that a biobank can take over the "custodianship" of biological materials and associated health data but does not become the legal owner of a collection, except the donor declares the abandonment of some or all his rights in the respective materials and data.[24] Similarly, the French National Ethics Committee (CCNE)[25] perceives of biobank managers as "guardians" (see Commin 2011) on the account that "the person who is at the source of the samples collected has rights which are not a form of ownership of the deposited elements" (2005, chap. 3). In contrast to this, the Greek National Bioethics Commission[26] allows for a transfer of ownership in samples to a biobank provided the donor's approval. According to the Greek Civil Code (art. 1061), biobanks may even obtain property in bodily materials if the way of how they have been processed qualifies for an acquisition of property.

Whilst for human tissue and biobank research also in the countries bordering the Mediterranean Sea informed consent is obligatory, in practice, however, several exemptions are granted. In France, for example, a semi-blanket consent is at place. According to this practice, the donors' consent is not restricted to a particular cancer research project, but may comprise any kind of research in cancer. Spain, similar to the French situation, follows a "flexible middle way" between open and specific consent, i.e. the donor's initial consent may include further unspecified uses. In fact, with the forthcoming Royal Decree, Spain is expected to switch to a blanket consent system (see Tiss.EU Report Paris 2009). Exemptions from consent are also granted for coded samples provided that its obtainment poses an unreasonable burden to researchers and that an Ethics Committee gives its approval. For a waiver of consent in the case of secondary research purposes, the Spanish BRA (art. 58.2) additionally requires that research occurs in the same institution. In Italy, in accordance with the *Authorisation of the Guarantor of Privacy of February 22nd* 2007, informed written

[24]See Paula Lobato de Faria's analysis presented at the Tiss.EU workshop in Paris, June 2009.

[25]Recommendation no. 77: ethical issues raised by collections of biological material and related data: "biobanks", "biolibraries" (2005).

[26]Recommendation on Banks of Biological Material of Human Origin (Biobanks) in Biomedicine Research (2006).

consent of the patient is obligatory in order to store biological samples and to manage genetic data, unless the data are used entirely for statistical or research purposes. For different research purposes, informed consent has to be obtained anew, but also here exemptions are granted provided that the samples and data are anonymised; that it is impossible to inform the patient despite reasonable efforts; and that the research program (which needs to be approved by the Ethics Committee) is also authorized by the Guarantor according to article 90 of the Data Protection Code. In the case of research and epistemiological studies, even the Portuguese *PGIHI* that actually provides for specific consent allows for exemptions. Although there is no discrete law on human tissue research in Malta, the current practice relies on a broad consent rule which is thus expected to become the default in a future legislation (see Tiss.EU Report Padova 2009).

In contrast to the former countries Greece seems to take a somewhat stricter stance on the consent issue. Both, the National Ethics Committee and the Greek *Data Protection Act*,[27] insist on the donor's free, informed and specific consent for the collection and processing of samples *and* data. Moreover, the *Act on Medical Research* even prohibits the obtainment of open or broad consent. Against this background, the Cyprian National Bioethics Commission (CyNBC)[28] seems to provide for a third way: Whilst for the collection of samples and data after 2004 the free and informed consent of the source is required, this consent can *either* be closed *or* open (2009, art. 13).

Data protection in the context of human tissue research is an important issue in the countries at the Mediterranean Sea's border. For example, the Italian National Bioethics Committee[29] requires the protection of the donor's privacy and anonymity. However, for research with anonymized samples, Greece, France, and Spain circumvent the necessity of consent because these are not subject to the countries' Data Protection Acts. Problems derive from the fact that, for example in Greece, anonymization is only vaguely defined ("if a person is no longer identifiable") (see Sándor et al. 2009). But also pseudonymisation – which is usually the favoured method for data protection by researchers – bears some pitfalls. For example in France, according to Fay Betsou, the police can break the code of a biobank sample via the clinician (see Tiss.EU Report Paris 2009). The most stringent regulations on data protection can be found in Spain and Portugal. In particular, the Portuguese PGIHI foresees the use of anonymized samples, except researchers are reliant on identifiable samples. In the latter case, coding is required whereby the codes need to be kept separately and only in public institutions. In Spain, biological samples have the same special protection status like personal health and genetic data. In particular, the BRA requires either consent or anonymization to use samples and health data in research.

[27] Act 2472/97 on the Protection of Individuals with Regard to the Processing of Personal Data implementing the European Directive 95/46/EC.

[28] Opinion on the Establishment and Use of Biobanks and Registries of Human Biological Samples for Research Purposes (2009).

[29] Biobanks and research on human biological material (2006).

Regarding the issue of feedback on incidental health-relevant findings two substantial approaches can be distinguished in this country group.[30] In France, on the one hand, information is only provided on overall research results, but no individual feedback is given to sample donors (see Commin 2011). On the other hand, in Spain, Cyprus, and Portugal donors are granted more rights in this respect. In particular, the Spanish BRA (art. 4.5) concedes donors a right to be informed of research results which might be relevant for their health and also a right to know their genetic data. In line with the CyNBC's opinon, the Cyprian *Law on the Processing of Personal Data* (sec. 12.6)[31] stipulates that "data relating to health shall be notified to the data subject through a doctor" and also the Portuguese *Act on the Protection of Personal Data* (art. 11.5) [32] grants the donor the right of access to information relating to personal health data, including genetic data, whereby this right is "exercised by means of the doctor chosen by the data subject."

In a nutshell, for the countries bordering the Mediterranean Sea it can be concluded that even in those States with rather strict consent requirements, several research facilitating exemptions are granted. Regarding the level of regulation it goes without saying that Portugal and Spain feature the most advanced frameworks in this group so far. However, the opinion of the Cyprian Bioethics Commission is remarkable for its sophisticated and comprehensive approach towards human tissue and biobank research. Whilst a differentiated regulation is most reasonable in this field – given the great variety of tissue research applications, it remains to be seen whether this can successfully be transposed into law.

10.3.2.7 The Eastern European Countries

As regards the existence of discrete jurisdictions on human tissue research, there is a sliding scale of progress in the Eastern European States: Whilst Hungary has adopted its *Act XXI on the Protection of Human Genetic Data, Human Genetic Tests and Research and Biobanks* ("Biobank Act") in 2008, Slovenia is expected to enact a *Law on Biomedical Research and Research on biological material of human origin* and Poland as well will regulate this issue in the near future (see Sándor, Śliwka, and Bárd, 2009). In Slovakia, the Czech Republic, Romania, and Bulgaria, by contrast, legal guidance can only be derived from laws covering related fields as well as from transnational codes and European documents. Notwithstanding these different levels of regulation, some common features in the countries' approaches to human tissue and biobank research can be identified.

The ethical and legal situation of the (South-) Eastern European countries is characterized by the joint experience of transition from the former regulation under

---

[30] This is irrespective of Malta, Greece and Italy where the existent frameworks remain silent on this issue.

[31] Protection of Individuals, Law 138 (I) 2001, amended 2003.

[32] See Act 67/98 which transposes the European Data Protection Directive 95/46/EC into Portuguese law.

communist influence to the necessity to adopt the Conventions and Directives of the European Union. Criticism was nourished by the impression that the adoption of the European *acquis communitaire* was not the free decision of the accession states' parliamentary bodies, but rather a tribute to international political necessities. In the context of research with human tissue and biobanking, there are three European documents in particular which had an impact on the regulation in this field: the Oviedo Convention, the EU Tissue and the EU Data Protection Directive. Whereas the two Directives had also to be transposed into national law by the Western European countries, the Oviedo Convention was only spatially inclusive and comprehensively ratified in the Eastern European states.

Given that the aforementioned documents follow specific normative prescriptions and principles, this naturally affects the regulation of property and ownership, informed consent as well as privacy and confidentiality (with the described difficulties, cf. Nys 2008). For example, for some Eastern European countries the transposition of European documents meant a de facto introduction of informed consent into the health care system. In particular, before Hungary changed its legislation in 2004, according to Zoltán Alexin "explicit written informed consent was nowhere used. Although patients had several veto rights, many times these were denied because the medical personnel was not aware of these rights, or vetoes could not be handled due to the organizational structure, or alternatively, the information systems were not designed to cope with vetoes" (Tiss.EU Report Budapest 2009, 12). As regards the acceptability of broad consent, rather diverging requirements can be found across this country group: whilst in the Czech Republic broad consent is not accepted for the removal of any body parts; the Slovenian Medical Ethics Committee recommends that patients may *either* give a blanket consent to all future uses of their stored specimens, *or* may opt to be asked for consent to any new use. In Hungary the question whether general consent is possible is not clearly answered by the Biobank Act (see Tiss.EU Report Budapest 2009).

Another characteristic feature of the Eastern European Countries is the predominance of opt-out-systems in therapeutic organ and tissue donation, which also affects the handling of research with tissue. For example, the Czech Republic applies a *presumed consent* model in the case of deceased persons, although informed consent is required to obtain samples from living persons (see Tiss.EU Report Budapest 2009, 10). Whilst also in Poland the Transplant Act of 2005 contains provisions concerning the establishment of tissue and cell biobanks, it is, however, not clear whether collections for research do also fall under the scope of this law.

Due to the Eastern European countries' historical experiences, the protection of privacy is highly valued in the post-communist era. For example, in Hungary due to concerns regarding genetic data protection other issues were left out from the Biobank Act (see Sándor and Bárd 2009a). Although in Poland there is a strong interest in privacy protection, too, it remains an open question whether besides data also samples fall under the Polish Data Protection Act as well (see Sándor, Śliwka, and Bárd 2009). Slovakia, in contrast, features a sophisticated coding system that does justice to both, traceability of samples *and* privacy protection (see Tiss.EU Report Budapest 2009).

Given their rather advanced legislations, Hungary and Slovenia are the only countries in this group that address the issue of feedback on health information. In particular, the Slovenian National Medical Ethics Committee recommends that information to source subjects on unexpected health-relevant findings must be provided together with appropriate counselling in a health care setting; the Hungarian Biobank Act similarly provides for a person's right to know the results of the human genetic study in a consultation that is specifically tailored to his or her needs but donors may also waive this right (see Sándor and Bárd 2009a, 6p).

10.4 Conclusions: Common Trends and Divergences for the Future Regulation and Harmonization of Human Tissue Research Within the European Union

Although our analysis of the seven European country groups was limited to four selected issues of human tissue and biobank research, the complexity of regulation is still remarkable at first sight. As this holds a danger of getting lost in details, for the conclusion we will step beyond this micro-perspective, and rather adopt a comprehensive view that allows for the identification of common trends but also remaining divergences.

To start with, regarding the legal status of human tissue samples, the majority of European Member States is most reluctant to grant donors property rights in their biological materials. Backed by the European documents' provision that the "human body and its parts shall not, as such, give rise to financial gain", both, the Continental Civil Law and the Common Law tradition, perceive of the human body as a *res extra commercium*. Despite this convergence in principle, however, the national jurisdictions also reveal some differentiation. In particular, we can distinguish between four approaches. Firstly, there are those countries whose jurisdictions do not take a stance at all towards this issue (e.g. Sweden), whilst others explicitly exclude any ownership or property rights in tissue samples and data derived thereof (e.g. France). Thirdly, few countries ascribe the right of ownership not to donors, but to the biobank (e.g. Greece) or to the chief processor of samples respectively (e.g. Estonia). The fourth group of countries sets out a de facto property right for donors, though bearing on diverging explanations: whilst on the one hand, de facto property rights are derived from the concept of personality rights (e.g. in the German speaking countries), other countries (e.g. Portugal) perceive of human genetic data as a common patrimony that is "owned" by everybody. These differences, however, ought not to distract from the existence of an additional similarity, namely that in the majority of countries the principle of non-commercialisation no longer applies if human body materials are turned into products (see Lenk and Beier 2011). Nonetheless, the criteria of this transition as well as its consequences for the legal status of processed materials are currently still a matter of controversy.

As regards the issue of informed consent, it is most remarkable that in several countries (e.g. the Eastern European countries, but also the UK), the necessity of regulating human tissue and biobank research in the first place triggered the legal enactment of informed consent procedures in health care and research. As a common denominator across the European Member States, the concept of informed consent in human therapy and research is widely acknowledged, whilst its limits (due to the peculiarities of human tissue and biobank research) are equally recognized. In light of these restrictions, there are tendencies towards a less strict interpretation of informed consent in practice (e.g. France). Evidence to this can be derived from the fact that countries with rather strict and specific consent regulation are currently considering a relaxation of these provisions (e.g. Sweden), while countries that are about regulating this field for the first time are expected to adopt a more liberal approach from the outset (e.g. Finland, Malta). The challenges of informed consent are also answered by reinterpretations of the consent rule. Several countries concede the possibility of broad/open consent (e.g. Estonia, Latvia, Switzerland, UK Biobank) which allows for the use of samples and data in immediate and future research projects of any kind at any time. In addition, the obtainment of consent for new research purposes is perceived as dispensable in a range of countries, provided that a Research Ethics Committee approves this kind of research (e.g. Sweden, Denmark, Spain, Lithuania), and/or the obtainment of secondary consent would entail an unreasonable effort compared to the expected benefits of research (e.g. Spain, Portugal). Other countries (for example Belgium) rely on an opt-out system for secondary uses of human tissues. However, it is important to note that the latter provisions conflict with the Oviedo Convention where it is stated that a secondary use is only acceptable provided the donor's informed consent (art. 22). On the other hand, the procedural approach which leaves the decision on informed consent to Research Ethics Committees is not unproblematic either. In particular, it needs to be legally defined who bears the ultimate responsibility.

Regarding the issue of anonymisation and pseudonymisation as means of protecting the donors' privacy, our analysis reveals that there is less heterogeneity in terminology and practice than typically assumed. In fact, the CoE's Recommendation provides for a quite useful definition. By distinguishing between identifiable and non-identifiable materials it avoids looming misconceptions of "anonymity" in the context of human tissue biobanking (art. 3). Whilst identifiable materials can either be coded, which implies that the code is accessible, the CoE speaks of "linked anonymised materials" if researchers cannot access the code. By defining non-identifiable materials as "unlinked anonymised materials" that "do not allow, with reasonable efforts, the identification of the person concerned," the CoE accounts for the impracticality of complete anonymisation in the era of genetic research. As a matter of fact, the CoE's definition is mirrored in several national legislations. Thereby, beyond terminological convergences, some more substantial commonalities can be identified. Firstly, the protection of samples and data in the context of biobank research is mostly – due to a lack of discrete laws on human tissue research – subject to national Data Protection Laws (as transpositions of the European Data Protection Directive), whereas only few countries make samples and data derived thereof subject to discrete laws (e.g. Sweden). Secondly, there is a tendency to exclude anonymised samples and data (i.e. those materials that cannot easily be traced back to the source's identity) from the protection measures that are otherwise applied to sensitive data. For example, in France, Greece, Spain, Finland, The Netherlands, Ireland but also Lithuania and Italy, the obtainment of consent is regarded dispensable in the case of research with anonymised samples. However, given that human tissue and biobank research is heavily dependent on the possibility of re-identifying donors, in the majority of European Member States coding is the preferred means for protecting the donors' privacy and personality rights. As regards the processing of coded samples, most countries require the separate storage of data and the key code (e.g. Portugal) and for the transfer of data and samples, it is mostly stipulated that the receiving institution or, the receiving country respectively, ought to adhere to the domestic rules applied in the sender country. Only Estonia excludes the transfer of samples altogether.

From our analysis we can furthermore conclude that the issue of feedback on incidental health findings is mainly addressed by countries featuring large-scale population biobanks and/or discrete human tissue and biobank research acts. Thereby the provision of feedback on health-relevant information is often solicited as a means to build a trust-relationship between donors and researchers (e.g. Estonia). On the basis of our analysis, four approaches towards the issue of feedback can be distinguished: Firstly, there are those countries which do not address feedback at all (e.g. France, most of the Eastern European States). Secondly, in some countries the national biobank projects exclude individual feedback (e.g. UK Biobank); or thirdly, the National Research Ethics Committee recommend it (e.g. Germany, Switzerland, Slovenia). Finally there are those countries where the provision of feedback is made obligatory by law (e.g. in the Baltic States, Belgium, Spain, Portugal, Hungary). Thereby the "right to know" is typically supplemented by a "right not to know". Moreover, it is also defined who is in charge of disclosing this information to the donor of the respective tissue. Given the importance of the donors' contribution to biobank research it is quite likely that the issue of individual feedback on health findings will be addressed by more and more national statutory documents in this field.

Taken together, it seems doubtful that a complete harmonization of all regulative issues concerning human tissue research will ever be attained. Given the countries' specific legal and ethical traditions it might even not be desirable, let alone feasible, to establish outright homogeneous norms in this field. Notwithstanding this result, our analysis revealed considerable convergences in several respects. This rapprochement can firstly be explained by the impact of the cited European documents which unfold a harmonizing effect particularly in the Eastern European States. Secondly, there is a cross-national learning noticeable: countries that regulated human tissue and biobank research more recently adopted features of already existing national legislations. A third incentive for harmonization of legal and ethical standards is intrinsic to human tissue and biobank research. In fact, the scientific value of biobanking increases if samples can be consolidated with other collections and even transferred abroad. For rendering this possible, almost all European Member States insist on protection measures that match their domestic provisions of sample and data protection. In the long run, this might therefore lead to additional adjustments – not only amongst the members of a country group, but even across the 27 members of the European Union, including Switzerland.

References

Anonymous. 2002. "Doctor Lied' to Parents over Autopsy Procedures." *The Guardian*. November 18, 2002. Accessed March 4, 2011. www.guardian.co.uk/uk/2002/nov/18/alderhey.

Beier, K. 2009. "Between Individualism and Solidarity: Biobanking in Sweden." In *New Challenges for Biobanks: Ethics, Law and Governance*, edited by K. Dierickx and P. Borry, 49–64. Antwerpen/Oxford/Portland: Intersentia.

Beier, K. 2011. "Beyond the Dichotomy of Individualism and Solidarity: Participation in Biobank Research in Sweden and Norway." In *Human Tissue Research – A Discussion of the Ethical and Legal Challenges from a European Perspective*, edited by C. Lenk, N. Hoppe, K. Beier, and C. Wiesemann, 65–75. Oxford: Oxford University Press.

Boggio, A., N. B. Adorno, E. Bernice, A. Mauron, and A. M. Capron. 2007. *Comparing Guidelines on Biobanks: Emerging Consensus and Unresolved Controversies*. Geneva, Switzerland: Réseau Universitaire International de Genève. Accessed June 14, 2011. http://www.ruig-gian.org/ressources/Boggio%20et%20al%20on%20Biobanks%20070827.pdf

Cambon-Thomsen, A., E. Rial-Sebbag, and B. M. Knoppers. 2007. "Trends in Ethical and Legal Frameworks for the Use of Human Biobanks." *European Respiratory Journal* 30 (2): 373–82.

Commin, V. 2011. "Legal Issues Surrounding French Research-Focused Biobanks." In *Human Tissue Research – A Discussion of the Ethical and Legal Challenges from a European Perspective*, edited by C. Lenk, N. Hoppe, K. Beier, and C. Wiesemann, 109–17. Oxford: Oxford University Press.

Dillner, J. 1999. "The Survey of Danish Biobanks Suggests a 'Drop-Out' Registry." *Läkartidningen* 99 (44): 4384–85.

Dillner, J. 2002. "Nya Biobankslagen antagen: Unik chance till förbättring" (New Act on Biobanks Passed: A Unique Chance for Improvement). *Läkartidningen* 99 (24): 2274–776.

Dörr, B. S. 2011. "Research with Human Biological Material and Personal Data in Biobanks – Legal and Regulatory Framework in Switzerland." In *Human Tissue Research – A Discussion of the Ethical and Legal Challenges from a European Perspective*, edited by C. Lenk, N. Hoppe, K. Beier, and C. Wiesemann, 99–107. Oxford: Oxford University Press.

Eensaar, R. 2008. "Estonia: Ups and Downs of a Biobank Project." In *Biobanks: Governance in Comparative Perspective*, edited by H. Gottweis and A. Petersen, 56–70. London and New York: Routledge.

Hansson, S. O., and B. Björkman. 2006. "Bioethics in Sweden." *Cambridge Quarterly of Health Care Ethics* 15: 285–93.

Johnston, C., and J. Kaye. 2004. "Does the UK Biobank Have a Legal Obligation to Feedback Individual Findings to Participants?" *Medical Law Review* 12: 239–67.

Kaye, J. 2006. "Do We Need a Uniform Regulatory System for Biobanks Across Europe?" *European Journal of Human Genetics* 14: 245–48.

Lenk, C. 2011. "Taking Solidarity Seriously: Do Biobank Institutions Have a Moral Obligation to Inform Their Patients on Incidental Health Findings?" In *Human Tissue Research – A Discussion of the Ethical and Legal Challenges from a European Perspective*, edited by C. Lenk, N. Hoppe, K. Beier, and C. Wiesemann, 55–63. Oxford: Oxford University Press.

Lenk, C., and K. Beier. 2011. "Is the Commercialization of Human Tissue and Body Material Forbidden in the Countries of the European Union?" *Journal of Medical Ethics* (forthcoming).

Madden, D. 2006. "Report of Dr. Deirdre Madden on Post Mortem Practices and Procedures." Government of Ireland, 2006. Accessed March 4, 2011. http://www.dohc.ie/publications/pdf/madden.pdf

Maschke, K. J. 2005. "Navigating an Ethical Patchwork – Human Gene Banks." *Nature Biotechnology* 23: 539–45.

Miola, J. 2011. "Law, Ethics and Human Tissue Research – Integration or Competition?" In *Human Tissue Research – A Discussion of the Ethical and Legal Challenges from a European Perspective*. edited by C. Lenk, N. Hoppe, K. Beier, and C. Wiesemann, 79–86. Oxford: Oxford University Press.

Nobel, S. 2008. Biobanks – Integration of Human Information to Improve Health. Report of the Committee for Research Infrastructures and the Scientific Council for Medicine at the Swedish Research Council. Accessed March 4, 2011. http://www.vr.se/download/18.61c03dad1180e26cb8780004896/Spr%C3%A5kgranskad+Biobank_report_SRC_v2.pdf.

Nys, H. 2008. "Research on Human Biological Materials and the Council of Europe: Some Unanswered Questions, Overlaps and Empty Boxes." *European Journal of Health Law* 15: 1–6.

Rynning, Elisabeth. 2009. "Legal Challenges and Strategies in the Regulation of Research Biobanking." In *The Ethics of Research Biobanking*, edited by Jan Helge Solbakk, Sören Holm, and Björn Hofmann, 277–314. Dordrecht/Heidelberg/London/New York: Springer.

Salter, B., and M. Jones. 2005. "Biobanks and Bioethics: The Politics of Legitimation." Accessed March 4, 2011. http://www.york.ac.uk/res/iht/projects/l218252005/SalterBiobanksAndBioethicsPaper.pdf

(SAMS) Swiss Academy of Medical Sciences. 2006. "Biobanks: Obtainment, Preservation and Use of Human Biological Material." Accessed June 14, 2011. http://www.samw.ch/de/Ethik/Richtlinien/Aktuell-gueltige-Rictlinien.html

Sándor, J., and P. Bárd. 2009a. *The Legal Regulation of Biobanks*. National Report: Hungary. CELAB Paper Series, No. 1. Accessed March 4, 2011. http://web.ceu.hu/celab/

Sándor, J., and P. Bárd. 2009b. *The Legal Regulation of Biobanks*. National Report: Latvia. CELAB Paper Series, No. 4.

Sándor, J., and P. Bárd. 2009c. *The Legal Regulation of Biobanks*. National Report: Estonia. CELAB Paper Series, No. 5.

Sándor, J., A. Drakopoulou, and P. Bárd. 2009. *The Legal Regulation of Biobanks*. National Report: Greece. CELAB Paper Series, No. 2.

Sándor, J., M. Śliwka, and P. Bárd. 2009. *The Legal Regulation of Biobanks*. National Report: Poland. CELAB Paper Series, No. 8.

Tiss.EU Report Birmingham. 2010. "Tiss.EU Workshop in Birmingham." June 3–4, 2009. Accessed March 4, 2011. www.tisseu.org.

Tiss.EU Report Budapest. 2009. "Tiss.EU Workshop in Budapest." April 6–8, 2009. Accessed March 4, 2011. www.tisseu.org.

Tiss.EU Report Hannover. 2008. "Tiss.EU Workshop in Hannover." November 28–29, 2008. Accessed March 4, 2011. www.tisseu.org.

Tiss.EU Report Padova. 2009. "Tiss.EU Workshop in Padova." September 24–26, 2009. Accessed March 4, 2011. www.tisseu.org.

Tiss.EU Report Paris. 2009. "Tiss.EU Workshop in Paris." June 24–25, 2009. Accessed March 4, 2011. www.tisseu.org.

Tiss.EU Report Stockholm. 2010. "Tiss.EU Workshop in Stockholm." March 24–26, 2010. Accessed June 14, 2011. www.tisseu.org.

UK Biobank Ethics and Governance Framework. 2007. Accessed June 14, 2011. http://www.ukbiobank.ac.uk/docs/EGFlatestJan20082.pdf

Ursin, L. Ø. et al. 2008. "The Informed Consenters: Governing Biobanks in Scandinavia." In *Biobanks: Governance in Comparative Perspective*, edited by H. Gottweis and A. Petersen, 177–193. London and New York: Routledge.

Chapter 11
Ireland and the United Kingdom's Approaches to Regulation of Research Involving Human Tissue

Elizabeth Yuko, Adam McAuley, and Bert Gordijn

11.1 Introduction

The Republic of Ireland (Ireland) and the United Kingdom (UK) share a common language, legal system, and membership of the European Union (EU) and Council of Europe (COE). However, there are also considerable cultural, social, economic, political, legal, religious, and moral differences between Ireland and the UK. These differences are reflected in Ireland and the UK's domestic regulation of research involving human tissue.[1] This chapter examines the extent to which international regulation affects a basic notion of political realism: that States take action based upon their national interests.[2] Before examining Irish and UK domestic regulation, it is necessary to discuss EU and COE regulation of research involving human tissue.

11.2 EU and COE Regulation of Research Involving Human Tissue

The EU and COE regulate research involving human tissue, but do so for different reasons and purposes which are reflected in the nature of their regulation. The EU's primary aim is economic integration; while the COE's primary aim is protection of human rights. Member States ceded greater sovereignty to achieve economic integration. EU law is supereme to national law and can be invoked by individuals in Member States. This is not to suggest that Member States cannot influence the drafting of EU law to protect their national interests.

[1]For the purposes of this chapter, residual embryos created following in vitro fertilisation (IVF) will be included alongside human tissue. The issue as to whether or not they are considered human tissue is debatable.

[2]See Donnelly (2000, 7).

E. Yuko (✉)
Institute of Ethics, Dublin City University, Dublin, Ireland
e-mail: elizabethyuko@gmail.com

C. Lenk et al. (eds.), *Biobanks and Tissue Research*, The International Library of Ethics, Law and Technology 8, DOI 10.1007/978-94-007-1673-5_11,

11.2.1 EU Tissue Directives

The original aim of the EU was to establish a single market in goods, capital, services, and persons. EU Member States added further to the EU's aims and increased the EU's powers. The new aims of the EU went beyond the establishment of the single market to include issues such as EU citizenship, foreign policy, and human rights. EU Member States conferred the EU with greater powers in relation to the EU's original and new aims. EU Member States ceded State sovereignty to protect public health in relation to human tissue and cells. This cession of State sovereignty was driven by the BSE/nvCJD (commonly referred to as Mad Cow Disease) crisis. This crisis demonstrated that the EU institutions needed greater competence and powers to protect public health in a single economic market.[3] Facing a public health scare, Member States understood the need for regulation at a supra-national level.

The Amsterdam Treaty strengthened the EU's role in the protection of public health by requiring a "high level" of human protection in the implementation of all European Community activities and policies. The Amsterdam Treaty permits the adoption of measures setting high standards of quality and safety of tissue and cells.[4] The EU adopted three directives: Directive 2004/23/EC, Directive 2006/17/EC and Directive 2006/86/EC ("EU Tissue Directives"). These EU Tissue Directives set the standards of quality and safety for the donation, procurement, storage and transfer of human tissues and cells.

A Directive is the legal device used to ensure that every Member State adopts a common or harmonised legal approach to an issue. A Member State is left to decide how to implement the Directive's requirements into national law. A Directive sets a time limit in which a Member State must implement the Directive's requirements. The European Commission may take legal action against a Member State who either fails to implement the Directive within the deadline or implements the Directive incorrectly.

The EU Tissue Directives regulate human tissue and cells that are applied to the human body in clinical trials,[5] but not tissue and cells used in in vitro research or animals.[6] Although there is only one reference in the preamble to tissue and cells applied to the human body in clinical trials,[7] such tissue and cells are regulated by the provisions of the EU Tissue Directives.

The EU Tissue Directives require Member States to designate a competent authority which is responsible for ensuring that human tissue and cells applied to the human body in clinical trials meet quality and safety standards;[8] ensuring

---

[3] See Hervey and McHale (2004, 77).

[4] Art. 152(4)(a).

[5] Directive 2004/23/EC (11).

[6] Ibid.

[7] Ibid.

[8] Ibid., Art. 4(1).

the appropriate supervision of human tissue and cell procurement;[9] the accreditation, designation, authorisation or licensing of tissue establishments and tissue and cell preparation processes;[10] inspections and control measures;[11] traceability;[12] import/export of human tissue and cells;[13] register of tissue establishments and reporting obligations;[14] notification of serious adverse events and reactions;[15] donor selection and evaluation;[16] provisions on the quality and safety of tissue and cells;[17] and exchange of information, reports and penalties.[18] The technical directives establish technical requirements for donation, procurement, and testing of human tissue and cells;[19] the traceability requirements, notification of serious adverse reactions, and events and certain technical requirements for the coding, processing, preservation, storage, and distribution of human tissue and cells.[20]

Although the primary aim of the EU Tissue Directives is to set standards for every aspect of the transfer, storage, and procurement of human cells and tissues, the directives contain extremely detailed provisions on the informed consent of donors and recipients which protect their human rights in terms of bodily integrity and dignity. However, the purpose of these protections is to ensure the functioning of the single economic market.

11.2.2 Council of Europe Oviedo Convention and Additional Protocols

A primary aim of the COE is the protection and strengthening of human rights. This is reflected in the Convention for the Protection of Human Rights and Dignity of the Human Being with regard to the Application of Biology and Medicine: Convention on Human Rights and Biomedicine ("Oviedo Convention"), which was adopted in 1997 and entered into force in 1999. It is also reflected in the four additional protocols to the Convention. Though only legally-binding on the States which have ratified the Convention, the Oviedo Convention has influenced the regulation of human tissue in a number of European States.

[9]Ibid., Art. 5.
[10]Ibid., Art. 6.
[11]Ibid., Art. 7.
[12]Ibid., Art. 8.
[13]Ibid., Art. 9.
[14]Ibid., Art. 10.
[15]Ibid., Art. 11.
[16]Ibid., chap. III.
[17]Ibid., chap. IV.
[18]Ibid., chap. V.
[19]Directive 2006/17/EC.
[20]Commission Directive 2006/86/EC.

EU Member States have ceded less sovereignty to the COE in relation to human tissue regulation. This is reflected in the fact that only 15 EU Member States have signed and ratified the Oviedo Convention.[21]

The Oviedo Convention does not specifically regulate human tissue for research purposes. It does, however, contain general provisions on scientific research. The Oviedo Convention requires that scientific research on human beings must be carried out freely, ensuring that the human being is protected and subject to the provisions of the convention or other legal regulations.[22] It also includes provisions on protecting participants who can and cannot consent to participation in research.[23] The Oviedo Convention requires adequate protection of the embryo, where a State permits research on embryos in vitro.[24] The Oviedo Convention prohibits the creation of embryos for research purposes.[25]

There are two additional protocols to the Oviedo Convention which are relevant to human tissue and cells. In 2002, an additional protocol was adopted on organ and tissue transplantation[26] for therapeutic purposes.[27] This protocol regulates professional standards;[28] health and safety;[29] organ and tissue removal from living[30] and deceased persons;[31] and prohibition on financial gain.[32] In 2005, an additional protocol was adopted on biomedical research, which regulates research activities in the health field involving interventions[33] on human beings.[34] It calls for the interests and welfare of the human being participating in the research to prevail over the sole

[21] Available at: http://www.coe.int/t/dg3/healthbioethic/source/INF(2010)1%20%C3%A9tat%20sign%20ratif%20r%C3%A9serves.doc (accessed 07 March 2011).

[22] Convention for the Protection of Human Rights and Dignity of the Human Being with regard to the Application of Biology and Medicine: Convention on Human Rights and Biomedicine, Art. 15.

[23] Convention on Human Rights and Biomedicine, Art. 16 and 17.

[24] Ibid., Art. 18(1).

[25] Ibid., Art. 18(2).

[26] The Protocol does not apply to reproductive organs and tissue; embryonic or foetal organs and tissues; or to blood and blood derivatives. Additional Protocol to the Convention on Human Rights and Biomedicine, on Transplantation of Organs and Tissues of Human Origin, Art. 3.

[27] Additional Protocol to the Convention on Human Rights and Biomedicine, on Transplantation of Organs and Tissues of Human Origin, Art. 2.

[28] Ibid., Art. 4.

[29] Ibid., Art. 6.

[30] Ibid., chap. III.

[31] Ibid., chap. IV.

[32] Ibid., chap. VI.

[33] For the purposes of this Protocol, the term "intervention" includes: a physical intervention, and any other intervention in so far as it involves a risk to the psychological health of the person concerned. Additional Protocol to the Convention on Human Rights and Biomedicine Concerning Biomedical Research, Art. 3.

[34] This Protocol does not apply to research on embryos in vitro. It does apply to research on foetuses and embryos in vivo. Ibid., Art. 2.

interest of society or science;[35] states that research should be carried out freely;[36] and that research on human beings may only be undertaken if there is no alternative of comparable effectiveness.[37]

Although the primary purpose of the Oviedo Convention and additional protocols is the protection of human rights, the Convention and additional protocols also seek to ensure safety in relation to the use of human tissue and cells.

11.3 Regulation of Research Involving Human Tissue and Cells in Ireland and UK

11.3.1 Ireland

The Irish legislature has failed to take the initiative to enact legislation that regulates research on human tissue and cells. The Minister for Health and Children did issue two statutory instruments in 2006 and 2007 in order to implement the EU Tissue Directives.[38] These statutory instruments offer the only regulation of research on human tissue and cells. Ireland implements the vast majority of EU directive by way of ministerial statutory instrument. The Irish Parliament does not discuss or vote to approve these statutory instruments.[39] These statutory instruments offer the only regulation of the use tissue and cells for research purposes.

In 2008, Senator Feargal Quinn, an independent senator, introduced a Private Member's Bill entitled the Human Body Organs and Human Tissue Bill, to regulate the removal, storage and use of human body organs and human tissue for certain purposes. A member of either house of the Irish legislature may place a Private Member's Bill before the house of which he or she is a member. The Government's legislative agenda determines the Government's attitude towards a Private Member's Bill. The Government will defeat such a Bill if the Bill's subject matter is not part of the Government's legislative agenda. The Government will welcome such a Bill if the Bill's subject matter is part of the Government's legislative agenda. The Government will ask that the member withdraw his or her Bill and undertake to publish its own Bill in the near future. In this case, the Deputy Minister for Health and Children indicated that the Minister for Health and Children was working on a

---

[35]Ibid., Art. 3.

[36]Subject to the provisions of this Protocol and the other legal provisions ensuring the protection of the human being. Ibid., Art. 4.

[37]Ibid., Art. 5.

[38]European Communities (Quality and Safety of Human Tissue and Cells) Regulations 2006 (SI No 158 of 2006) transposes Directive 2004/23/EC and Directive 2006/17/EC; European Communities (Human Tissue and Cells Traceability Requirements, Notification of Serious Adverse Reactions and Events and Certain Technical Requirements) Regulations 2007 (SI No 598 of 2007) transposes Directive 2006/86/EC.

[39]See Byrne and McCutcheon (2001, 442–43) for further information on Irish statutory instruments.

Human Tissue Bill, but did not want to make any decisions – particularly regarding organ donation for transplantation – without first consulting the public.[40]

On 9 April 2009, the Minister asked for the public's views on an extremely detailed draft proposal[41] for a Human Tissue Bill.[42] While a summary of these views were made publicly available in August 2009, no further action has yet been taken on either the Private Member's Bill, or the Minister's Bill. Indeed, the Minister never published this Bill. If the current Government ever publishes this Bill, it will probably be similar to the detailed proposal.

Many of the Draft Proposals' provisions are drawn from the recommendations contained in Dr. Deirdre Madden's Report on organ removal and retention arising from paediatric post-mortems ("Madden Report").[43] For example, the Draft Proposals' proposition that authorisation is required for hospital post-mortem examination and tissue retention is drawn directly from the Madden Report's recommendations.[44] The Draft Proposals also regulate other matters which were not considered in the Madden Report, such as consent procedures for the use of tissue from living and deceased donors for research purposes.[45]

The Draft Proposals place primacy on consent, authorization and autonomy. This is reflected in two of its guiding principles. Firstly, the bodily integrity of the individual before and after death must be protected. Secondly, the autonomy of the individual and the rights of the bereaved must be respected. It is also reflected in the Draft Proposals' provisions regulating the use of human tissue taken from the living and deceased for research purposes. For example, consent for research on organs and tissue must be sought, even though consent for the removal and retention of organs and tissue was previously obtained as part of a post-mortem process.[46] Tissue from the deceased may be used for research where this was authorized by an advance healthcare directive, by next-of-kin, or a nominated proxy.[47] This consent may be general, specific, limited, or qualified.[48] Furthermore, subsequent research on the same tissue must be submitted for approval if the research is substantially different in nature to the approval originally given.[49] The Draft Proposals also regulate research on any tissue obtained in the past. Such tissue may be used where the

[40] See http://historical-debates.oireachtas.ie/S/0191/S.0191.200810010008.html (accessed 07 March 2011).

[41] The Department of Health and Children (2009). Available at: http://www.dohc.ie/consultations/closed/human_tissue_bill/draft_proposals.pdf?direct=1 (accessed 07 March 2011).

[42] See http://www.dohc.ie/consultations/closed/human_tissue_bill/covering_letter.pdf?direct=1 (accessed 07 March 2011).

[43] Madden (2006b). Available at: http://www.dohc.ie/publications/madden.html (accessed 07 March 2011).

[44] The Department of Health and Children (2009, 6).

[45] Ibid., 6–7.

[46] Ibid., 81.

[47] Ibid., 89–97.

[48] Ibid., 129.

[49] Ibid.

researcher cannot identify the person from whom the tissue was taken and a research ethics committee approves the research; or where the tissue has been imported.[50]

The Draft Proposals permit research on tissue from living donors where consent is given freely and without coercion; without the promise of benefits likely to result from participation; and on the basis of appropriate information on the nature and purpose of the research.[51] If tissue from a living person was donated as a by-product of a medical procedure, consent for future unspecified uses may be obtained.[52] A donor may freely withdraw authorization or consent at any time.[53]

Embryonic stem cell research is a controversial subject in Ireland, which explains why the definition of "tissue" in the Draft Proposals does not include human gametes and embryos.[54] Until recently, it was widely presumed that the constitutional protection of "the right to life of the unborn"[55] prohibited research on embryos[56] and pre-empted many of the related legal and ethical dilemmas which require regulation by statute.[57] In *Roche v. Roche*[58] a unanimous Irish Supreme Court found that this constitutional provision did not extend protection to frozen embryos because the sole purpose of this provision was to prevent abortion. Four members of the Supreme Court examined whether any other provision of the Constitution offered protection to frozen embryos. Chief Justice Murray did not exclude the possibility that a frozen embryo may be protected by Article 40.3 whereby the State guarantees in its laws to respect, and, as far as practicable, by its laws to defend and vindicate a citizen's personal rights. Article 40.3 refers to the rights to "life, person, good name and property."[59] The difficulty with the Chief Justice's approach is that the rights to good name and property cannot apply to an embryo, and applying the rights to life and person to the embryo suggests that the embryo may have rights to life and physical integrity. Chief Justice Murray noted that the human embryo has moral qualities and status. It contains the potential for life and cannot be divorced from concepts of human dignity. Chief Justice Murray believed that the ethical and moral status of the embryo as inextricably linked with the concept of human dignity. The Chief Justice drew support for this proposition from three sources of international law prohibiting the creation of human embryos for research and cloning: the Oviedo Convention,[60]

[50]Ibid.

[51]Ibid., 132.

[52]Ibid.

[53]Ibid.

[54]Ibid., 14.

[55]Constitution of Ireland, Art. 40.3.3.

[56]Madden (2006a, 33).

[57]McDonnell and Allison (2006, 818).

[58]Roche v.Roche & ors, Judgment of Mr. Justice Murray. (2009) IESC 82, 15 December 2009.

[59]Constitution of Ireland, Art. 40.3.2.

[60]Convention for the Protection of Human Rights and Dignity of the Human Being with regard to the Application of Biology and Medicine: Convention on Human Rights and Biomedicine.

the EU's Charter of Fundamental Rights[61] and a United Nations Declaration on Human Cloning.[62]

The three remaining members of the Supreme Court did not invoke Article 40.3. Justice Hardiman doubted that the unborn was entitled to constitutional protection under Article 40.3 but found that it "does not of course mean that such embryos should not be treated with respect as entities having the potential to become a life in being."[63] Justice Geoghegan also believed that spare embryos ought to be treated with respect.[64] Justice Geoghegan held that the absence of statutory regulation providing this respect was undesirable and arguably contrary to the spirit of the Constitution.[65] Justice Fennelly agreed with Justice Hardiman and Justice Geoghegan that the embryo is entitled to respect.[66] Justice Fennelly believed that there might be a constitutional obligation on the State to give concrete form to that respect.[67] Justice Fennelly stated that it may be open to a court in a future case to consider whether an embryo enjoys protection under other provisions of the Constitution, where the executive and legislative organs of the State had failed to take action.[68] Therefore, the Irish legislature will have to include frozen embryos in the Human Tissue Bill's definition of tissue and a prohibition on any form of embryonic research that threatens embryo viability.

The Irish Medical Council's *Guide to Professional Conduct and Ethics for Registered Medical Practitioners* also provides indirect regulation of the use of human embryos for research purposes. The Guide prohibits creating new forms of life solely for experimental purposes.[69] The Medical Council may remove a medical practitioner from the register for professional misconduct where the medical practitioner breaches the guide. However, the Guide does not regulate scientists who work with embryos such as embryologists, and this demonstrates the need for research on embryos to be regulated by human tissue legislation.

11.3.2 UK

The regulatory landscape of the UK stands in stark contrast to that of Ireland. The UK has comprehensive regulation of human tissue. The Human Tissue Act 2004 and

---

61 The Charter of Fundamental Rights of the European Union (2000). Available at: http://www.europarl.europa.eu/charter/default_en.htm (accessed 07 March 2011).

62 UNGA Resolution 59/280 (2005). Available at: http://www.un.org/law/cloning/ (accessed 07 March 2011).

63 Roche v. Roche & ors, Judgment of Mr. Justice Hardiman (2009) IESC 82, 15 December 2009.

64 Roche v. Roche & ors, Judgment of Mr Justice Geoghegan (2009) IESC 82, 15 December 2009.

65 Roche v. Roche & ors, Judgment of Mr Justice Geoghegan (2009) IESC 82, 15 December 2009.

66 Roche v. Roche & ors, Judgment of Mr Justice Fennelly. (2009) IESC 82, 15 December 2009; Roche v. Roche & ors, Judgment of Mr Justice Geoghegan. (2009) IESC 82, 15 December 2009; Roche v. Roche & ors, Judgment of Mr Justice Hardiman. (2009) IESC 82, 15 December 2009.

67 Roche v. Roche & ors, Judgment of Mr Justice Fennelly. (2009) IESC 82, 15 December 2009.

68 Roche v. Roche & ors, Judgment of Mr Justice Fennelly. (2009) IESC 82, 15 December 2009.

69 Irish Medical Council (2009).

the Human Tissue (Scotland) Act 2006 regulate human tissue for research purposes in England, Wales, Northern Ireland and Scotland. The Human Tissue Act 2004 has a wider scope than its Scottish equivalent. Like the Irish Draft Proposals, the Human Tissue Act 2004 was drafted in response to organ retention scandals involving the storage of thousands of organs, tissue samples, body parts, stillbirths, and foetuses without consent.[70] It repealed the Human Tissue Act 1961, the Anatomy Act 1984, and the Human Organ Transplant Act 1989.[71]

The Human Tissue Act 2004 covers the storage, use, and removal of human tissue.[72] It established the Human Tissue Authority (HTA).[73] The 2004 Act regulates the storage and use of human tissue for the purpose of research in connection with disorders, or the functioning of the human body.[74]

The 2004 Act places primacy on consent. Human tissue from living persons may be used for research purposes where the "appropriate consent" has been obtained. The nature of "appropriate consent" varies depending from whom tissue was obtained.[75] Bodies of the deceased may be stored[76] and used[77] for research purposes, and tissue may be removed[78] from a deceased's body for research purposes, where appropriate consent has been obtained. In addition, the Secretary of State may make regulations that allow the English High Court to make an order deeming that consent would be granted for the storage, use, or removal of the body of a deceased person, or relevant material[79] from the body of a deceased person; and for the storage and use of relevant material which has come from a living person, for research purposes.[80]

The Human Tissue (Scotland) Act 2006 focuses on the material removed from deceased bodies and does not cover the use or storage of tissue from living individuals, with the exception of the purpose of transplantation.[81] The HTA is also the competent authority for Scotland.[82] Tissue from deceased persons[83] and tissue removed during a post-mortem examination[84] may be removed and used for research purposes with appropriate authorisation.

---

[70]See McHale (2005, 169–71) and Price (2005, 798).

[71]McHale (2005, 171–72).

[72]Human Tissue Act 2004, Art. 1.

[73]Ibid., Art. 13(1).

[74]Ibid., Schedule 1, Part 1.

[75]Ibid., Art. 1(1), (7), (8), (9) (10).

[76]Ibid., Art. 1(1)(a).

[77]Ibid., Art. 1(1)(b).

[78]Ibid., Art. 1(1)(c).

[79]"Relevant material" means material, other than gametes, which consists of or includes human cells. Human Tissue Act 2004, Art. 53(1).

[80]Human Tissue Act 2004, Art .7(4).

[81]Human Tissue (Scotland) Act 2006, Art. 3 and 17.

[82]Ibid., Art. 54(1).

[83]Ibid., Art. 3. In the Human Tissue (Scotland) Act 2006, the word "authorisation" is used instead of "consent."

[84]Ibid., Art. 28.

While the Human Tissue Act 2004 regulates many of the EU Tissue Directives' requirements to varying extents, the 2004 Act preceded the EU Tissue Directives. Like Ireland, the UK implemented the EU Tissue Directives by way of ministerial statutory instrument. The Human Tissue (Quality and Safety for Human Application) Regulations 2007 provide the necessary adjustments to the existing regulatory framework to comply with the EU Tissue Directives' requirements, as well as extending certain provisions to Scotland.[85] The statutory instrument applies to England, Northern Ireland, Wales, and Scotland.[86] It is largely technical, providing regulations on licensing; duties of the HTA in relation to tissue and cells; and amendments to the Human Tissue Act 2004.[87]

The UK also passed the Human Fertilisation and Embryology Act 2008 to make amendments to the Human Fertilisation and Embryology Act 1990 in order to comply with the EU Tissue Directive's provisions regulating human reproductive tissues and cells. However, the 2008 Act goes further than the EU Tissue Directives by regulating the use of embryos for research purposes.[88] Embryos may be created in vitro, kept and used for the purposes of a research project specified in the licence.[89] This also applies to human admixed[90] embryos.[91] A license to conduct research on embryos can be obtained for a list of "principal purposes" set out in the Act, such as increasing knowledge about or developing treatments for serious diseases or other serious medical conditions;[92] promoting advances in the treatment of infertility;[93] increasing knowledge about the causes of miscarriage;[94] developing more effective techniques of contraception;[95] developing methods for detecting the presence of gene, chromosome, or mitochondrion abnormalities in embryos before implantation;[96] and increasing knowledge about the development of embryos.[97]

---

[85] Statutory Instrument No. 1523, The Human Tissue (Quality and Safety for Human Application) Regulations 2007.

[86] Statutory Instrument No. 1523, The Human Tissue (Quality and Safety for Human Application) Regulations 2007, Art. 1.

[87] Statutory Instrument No. 1523, The Human Tissue (Quality and Safety for Human Application) Regulations 2007.

[88] Human Fertilisation and Embryology Act 2008, Schedule 2(6).

[89] Human Fertilisation and Embryology Act 2008, Schedule 2(6)(1).

[90] In this Act, an "admixed embryo" is an embryo created by using a mixture of human and animal gametes, pronuclei or parts of each; or a human embryo that has been altered by the introduction of animal cells or DNA. Human Fertilisation and Embryology Act 2008, Art 4(6).

[91] Human Fertilisation and Embryology Act 2008, Schedule 2(6)(3).

[92] Ibid., Schedule 2(6)(3A)(2)(a),(b).

[93] Ibid., Schedule 2(6)(3A)(2)(d).

[94] Ibid., Schedule 2(6)(3A)(2)(e).

[95] Ibid., Schedule 2(6)(3A)(2)(f).

[96] Ibid., Schedule 2(6)(3A)(2)(g).

[97] Ibid., Schedule 2(6)(3A)(2)(h).

11.3.3 Shared Similarities and Experiences

Ireland and the UK share similarities and experiences. First, Ireland and the UK each have a common law system with a parliament which is charged with drafting legislation. The Irish Parliament has replicated the British Parliament's legislative approaches to health care regulation. For example, Irish and UK legislation on criminalisation of assisted suicide and consent to treatment of children over 16 is identical,[98] albeit Ireland introduced these laws some 30 years after their introduction in the UK. Both the Irish Draft Proposals and English Human Tissue Act 2004 place primacy on free and informed consent. The Irish Parliament, however, has been slow to regulate many aspects of health care such as advance care directives, abortion, and mental health and capacity.

Secondly, Ireland has a written constitution which contains express and implied human rights relevant to health care,[99] such as bodily integrity, privacy, autonomy, and dignity.[100] The UK's constitutional system contains a small number of human rights. However, Ireland and the UK share a common human rights framework because both are signatories[101] to the European Convention on Human Rights. The UK incorporated this convention into its domestic law with the passing of the Human Rights Act 1998[102] and Ireland with the passing of the European Convention on Human Rights Act 2003.[103] Both States consider that respect for bodily integrity, consent, and autonomy are fundamental to research on human tissue, which is reflected in the English Human Tissue Act 2004 and the Irish Draft Proposals.

Thirdly, Ireland and the UK experienced organ retention scandals and established inquiries into these scandals.[104] However, each State had a different regulatory response. The UK introduced comprehensive human tissue regulation in light of the public reaction to the organ retention scandals. Ireland introduced limited regulation of tissue and cells in order to comply with Ireland's EU legal obligations.

Fourthly, Ireland and the UK have refused to sign the Oviedo Convention because of the Convention's regulation of embryos for research. Ireland found these provisions to be too liberal, permitting the destruction of human embryos. Former Minister of State Ivor Callelly stated that "there are difficulties with a number of

[98]See Ireland – Criminal Law (Suicide) Act 1993, Sec. 2 and UK – Suicide Act 1961, Sec. 2. Ireland – Non-Fatal Offences Against the Person Act 1997, Sec. 23 and UK – Family Law Act 1969, Sec. 8.

[99]See Byrne and McCutcheon (2001, 52); Constitution of Ireland 1937.

[100]The Constitution Review Group (1996, 221–22); Constitution of Ireland 1937, Art. 40.3.

[101]For list of signatories, see http://conventions.coe.int/Treaty/Commun/ListeTableauCourt.asp?MA=3&CM=16&CL=ENG (accessed 08 March 2011).

[102]Human Rights Act, 1998. Available at: http://www.opsi.gov.uk/acts/acts1998/ukpga_19980042_en_1#pb1-l1g1 (accessed 08 March 2011).

[103]European Convention on Human Rights Act, 2003. Available at: http://www.irishstatutebook.ie/2003/en/act/pub/0020/index.html (accessed 08 March 2011).

[104]See http://www.dohc.ie/publications/pdf/madden.pdf?direct=1 for the Madden Report from Ireland, and http://www.rlcinquiry.org.uk/ for the Redfern Report from the UK (accessed 08 March 2011).

[Oviedo] articles that have implications for the destruction of human embryos."[105] The UK's reason for not signing the Oviedo Convention contrasts sharply with that of Ireland. The UK found these provisions to be too conservative, which would inhibit research. Since 1990, UK law has permitted research on residual embryos from IVF treatment for assisted reproduction, as well as embryos specifically created for research purposes.[106] The Oviedo Convention prohibits the creation of human embryos for research purposes.[107] The UK would have to enter a reservation in relation to this prohibition of the Oviedo Convention in order to sign and ratify the Oviedo Convention.

11.4 Explanation of Differences

Isasi and Knoppers found that legal traditions, cultural and socio-religious beliefs, and economic interests inform and shape public policy on controversial issues such as embryonic, cloning and stem cell research (Isasi and Knoppers 2006).[108] Ireland's approach to these issues has been described as restrictive[109] while the UK's approach has been described as liberal (Isasi and Knoppers 2006).[110] It is possible to identify reasons which explain these differences.

11.4.1 Bioethics, Religion, and Regulation

Dickenson identifies four models or "voices" of European bioethics (Dickenson 1999).[111] Although geographically located in Western Europe, Dickenson believes that Ireland should be included with Southern European countries which have rejected the liberal rights-oriented model of the person adopted by the UK and other Western European States, in order to enshrine a positive duty of seeking to promote one's own health in its constitution (Dickenson 1999).[112] This is significant because those classifications directly reflect the bioethics and regulatory approaches of Ireland and the UK. Dickenson identifies the influence of the Catholic Church as the common tie between Ireland and the countries of Southern Europe (Dickenson

---

[105] Seanad Éireann. Volume 170. 4 December 2002. Adjournment Matters. Oviedo Convention.

[106] Walters (2004, 5).

[107] Convention for the Protection of Human Rights and Dignity of the Human Being with regard to the Application of Biology and Medicine: Convention on Human Rights and Biomedicine, Art. 18(2).

[108] Isasi and Bartha (2006, 16–17).

[109] Ibid., 20; Caulfield (2003, 88).

[110] Isasi and Bartha (2006, 22–23), Caulfield (2003, 89).

[111] Dickenson (1999, 249).

[112] Ibid., 255.

1999).[113] She notes that despite the fact that Ireland has never been under Fascist rule – like the countries of Southern Europe – it has experienced an enduring absence of pluralism (Dickenson 1999).[114]

Gunning and Caulfield also found that a State's prohibition on embryo research was caused primarily, but not entirely, where the Roman Catholic Church's influence is strong (Gunning 1999 and Caulfield 2003).[115] Barrington supported this proposition by suggesting that the root causes for the dearth of Irish regulation are the dominance of the Catholic Church and the non-existent public debate on scientific developments (Barrington 2002).[116]

However, as Ireland has become more multicultural in the twenty-first century, there has not yet been a discernible shift to a more liberal approach to bioethical issues in terms of regulation. Two factors have diluted the influence of the Roman Catholic Church in Ireland. First, there has been the arrival of a large number of immigrants with different cultural and religious backgrounds. Secondly, many Irish people have moved away from strict adherence to the tenets of the Roman Catholic Church. For example, there has been a considerable increase in the number of couples who cohabit rather than marry.[117] Furthermore, other countries in Southern Europe, such as Italy and Spain, have enacted some type of law regulating assisted human reproduction, which Ireland has yet to do. As a result, Catholicism is not the sole reason for the absence of laws in Ireland to regulate this research.

During negotiations for the EU Seventh Framework Programme's funding of embryonic stem cell research, Irish politicians, like the Irish population, were divided.[118] Ultimately, those representing Ireland believed that they should not prevent such research from being undertaken in other Member States regardless of whether such research should ever occur in Ireland.[119] This suggests either a weakening influence of the Catholic Church in Ireland in terms of determining a bioethical position, or "a prioritisation of the economic interests of medical research over traditional ethical values."[120]

Currently, religion, bioethics, and regulation in Ireland reflect one particular moral or bioethical view which is dominated by a Roman Catholic ethos. The introduction of any comprehensive regulation would involve a debate on the different bioethical positions which would underpin such regulation, which, of course, Irish politicians would be reluctant to do. This allows the perceived dominant conservative bioethical position to remain unchallenged.

---

[113] Ibid.

[114] Ibid.

[115] Gunning (1999, 166) and Caulfield (2003, 88).

[116] Barrington (2002, 147).

[117] Irish Census 2006 (2007, 102–06).

[118] Madden (2006a, 33).

[119] Ibid.

[120] Ibid.

The issue is whether the public support or oppose this conservative bioethical perspective. Despite the fact that the Church has fewer adherents, the public's stance on many social – and bioethical – issues may still clearly reflect those of the Catholic Church. In 2005, a survey sought to measure public approval and disapproval of four types of biomedical research. Of the four types of research, the highest approval was for stem cell research using adult human tissue (49%) and cloning human cells to combat disease (42%).[121] Conversely, the study found that the highest levels of disapproval were for the development of genetically modified foods (52%) and stem cell research using human embryos (34%).[122] However, a significant minority, ranging from one-quarter to one-third of the participants, were undecided about each of these issues.[123]

Also of note, 42% of people surveyed believe that human genetic research was tampering with nature, while one-third disagreed, and 24% were unsure.[124] However, one-third of participants in a similar survey conducted in the UK believe that human genetic research was tampering with nature, suggesting that the Irish public is slightly more negative towards genetic research than their UK counterparts.[125]

A lack of political will or courage is the reason for the absence of comprehensive human tissue regulation in Ireland. The only regulation that has been adopted in this area was required because of Ireland's EU membership. It did not originate in Ireland. Irish politicians fear initiating or supporting bioethics-related regulation may harm their electability which is reflected in the failure to publish let alone adopt the Minister's Bill.

While there are few votes to be gained supporting human tissue regulation in Ireland, a strong vocal minority may ensure that members of parliament lose votes by supporting such regulation. The politicians may also believe that the electorate cannot distinguish between regulations of research on human tissue, embryonic stem cell research and abortion. Politicians may be weary and wary of revisiting the divisive debates on abortion caused by referenda and high-profile court cases. This is reflected by the fact that two bioethics-related bills were introduced by members of the Seanad as Private Member's Bills: the Regulation of Assisted Human Reproduction Bill 1999[126] and the Human Tissue Bill 2008.[127] These senators are independent and represent constituencies whose electorate consists solely of university graduates. These two factors allowed these senators to propose "controversial"

[121] Cousins et al. (2005, 38).

[122] Ibid.

[123] Ibid.

[124] Ibid.

[125] Ibid.

[126] See http://www.oireachtas.ie/viewdoc.asp?DocID=2365&&CatID=59&StartDate=01%20January%201999&OrderAscending=0 (accessed 08 March 2011).

[127] See http://www.oireachtas.ie/viewdoc.asp?fn=/documents/bills28/bills/2008/4308/document1.htm (accessed 08 March 2011).

legislation without risking their seats. Such a risk exists for a member of parliament who is a member of a political party and represents a geographic constituency comprised of voters on the general electoral register.

However, the results of a small public opinion survey indicate that the Irish public may not be as conservative as politicians think. For example, 82% of people surveyed said that surplus embryos should be used for medical research into disease, even if that means they will be destroyed.[128] Furthermore, 53% of those surveyed either agree or strongly agree that the Irish Government should provide funding for embryonic stem cell research.[129]

The position and role of the established Church of England is different where several prominent members of the Church of England have taken a liberal stance. Furthermore, the bishops are constitutionally entitled to a seat in the House of Lords and may influence the framing of legislation.[130] In the 1990s, Archbishop of York John Hapgood supported legislation in favour of embryo research.[131] Ten years later, the Bishop of Oxford, who sat in the House of Lords, supported amendments to the Human Fertilisation and Embryology Act in favour of stem cell research.[132] Mulkay suggests that opposition to embryonic stem cell research in the UK was weakened by inconsistent religious leadership and by divisions of religious opinion, "which made religious condemnation of research on human embryos appear increasingly arbitrary" (Mulkay 1997).[133]

11.4.2 Science and Economics

The UK has a strong scientific infrastructure which is reflected in the UK being at the forefront of many scientific developments, such as IVF and therapeutic cloning. This explains the UK's policies are the most liberal policies on human embryonic stem cell research in Europe. Plomer argues that ultimately it was the anticipated benefits from research and the social and cultural authority of the scientific community that allowed the Human Fertilisation and Embryology Act to pass in 1990 (Plomer 2002).[134]

McHale contends that amendments to the UK's Human Tissue Act when it was a Bill may represent a shift away from respect for individual autonomy towards respect for the researcher (McHale 2005).[135] She uses the example of concerns over the Bill's provisions enabling the use of anonymous "spare" material without

[128] Irish Council for Bioethics (2005, 34).

[129] Irish Council for Bioethics (2005, 36).

[130] Plomer (2002, 137).

[131] Ibid.

[132] Ibid.

[133] Mulkay (1997) quoted in Plomer (2002, 138).

[134] Plomer (2002, 138).

[135] McHale (2005, 186).

consent.[136] Price states that the Government was "undoubtedly exercised" by the prospect of researchers seeking opportunities abroad if the use of human tissue was overregulated in the UK, therefore jeopardising Britain's status as a major research player (Price 2005).[137] The UK's scientific and medical community had significant influence on the development of the Human Tissue Act 2004. The importance of science should not be underestimated. This is reflected in Sweden and Finland's regulations of research on in vitro embryos. These regulations result from the interplay of politics, economics and history of past accomplishments in scientific research. Sweden and Finland have a strong academic tradition of developmental biology, and in vitro embryo and stem cell research is a logical continuum to this history. The attitudes to science in both of these countries are also highly positive, and investments in biotechnology could even be described as a national project.[138]

Ireland does not have a strong history or the same level of financial investment in this form of biomedical research as the UK. However, there is no way of knowing the extent to which research is – or is not – taking place in Ireland where such research is unregulated.

11.5 Conclusions

A State's domestic concerns reflect the existence or non-existence of domestic regulation and the nature of domestic regulation. Despite the tissue retention scandals, Ireland does not have comprehensive regulation of human tissue because the Government is concerned that the public or a vocal minority may oppose regulation believing that this will lead to the destruction of frozen embryos. The UK decided to regulate human tissue because of the public disquiet caused by the tissue retention scandals, recognition of the financial benefits of such research and the fact that religious and public opinion was liberal.

The EU Tissue Directives and Oviedo Convention demonstrate that the success or failure of European regulation depends on the willingness to accommodate States' domestic concerns. The EU Tissue Directives' regulatory system is concerned with safety and protecting the interests of recipients and donors. Ireland and UK share this concern for recipient and donor safety and informed consent.

More importantly, the EU Tissue Directives decline to regulate areas where there is little or no moral or ethical consensus amongst the Member States such as in vitro research on human tissue and cells, the use of germ and embryonic stem cells and the legal definition of "person" or "individual." Ireland and the UK welcome this omission but for different domestic concerns. Ireland could not agree to allow biomedical research involving human embryos because of the public opposition to such research. The UK could not agree to allow restrictions to be placed

---

[136] Ibid.

[137] Price (2005, 819).

[138] See Walin (2007, 147).

on biomedical research involving embryos because of the strong scientific community opposition to such restrictions. The Oviedo Convention regulates the human embryo, and this is one reason why the Oviedo Convention remains European regulation in embryonic form. This is reflected in the limited number of ratifications by European States and failure to agree to an additional protocol on embryonic protection.[139] Ireland and the UK have neither signed nor ratified the Oviedo Convention, again for different domestic concerns, because Ireland believes that it is too liberal while the UK believes it is too conservative.

EU regulation in this area may lead to further homogenization in Ireland and the UK. However, Ireland and the UK will probably retain their national peculiarities, particularly in relation to research involving human tissue.

The future may see a gradual and subtle shift in Ireland to more liberal human tissue policies. For example, scientists may be permitted to perform research on imported stem cell lines because no embryos are created or destroyed in Ireland. In 2008, University College Cork was the first university to permit such research. Furthermore, Ireland's talk of a knowledge economy and the importance of biotechnology may be liberalising influences, particularly during this economic recession. The UK, however, is likely to retain its current level of regulation and commitment to research.

11.6 Sources of Legal Regulation

Constitution of Ireland 1937.
Suicide Act 1961.
Family Law Act 1969.
Criminal Law (Suicide) Act 1993.
Non-Fatal Offences Against the Person Act 1997.
Human Rights Act 1998.
Convention for the Protection of Human Rights and Dignity of the Human Being with regard to the Application of Biology and Medicine: Convention on Human Rights and Biomedicine 1999.
Treaty of Amsterdam 1999.
The Charter of Fundamental Rights of the European Union 2000.
Additional Protocol to the Convention on Human Rights and Biomedicine, on Transplantation of Organs and Tissues of Human Origin 2002.
European Convention on Human Rights Act 2003.
Directive 2004/23/EC.
Human Tissue Act 2004.
Additional Protocol to the Convention on Human Rights and Biomedicine Concerning Biomedical Research 2005.
UNGA Resolution 59/280 2005.

[139] See: Braake (2004, 144), Walin (2007, 158).

Directive 2006/17/EC.
Directive 2006/86/EC.
European Communities (Quality and Safety of Human Tissue and Cells) Regulations 2006 (SI No 158 of 2006).
Human Tissue (Scotland) Act 2006.
European Communities (Human Tissue and Cells Traceability Requirements, Notification of Serious Adverse Reactions and Events and Certain Technical Requirements) Regulations 2007 (SI No 598 of 2007).
Statutory Instrument No. 1523, The Human Tissue (Quality and Safety for Human Application) Regulations 2007.
Human Fertilisation and Embryology Act 2008.
Roche v. Roche & ors, Judgment of Mr. Justice Hardiman (2009) IESC 82, 15 December 2009.
Roche v. Roche & ors, Judgment of Mr Justice Murray. (2009) IESC 82, 15 December 2009.

References

Barrington, R. 2002. "Terrible Beauty or Celtic Mouse? The Research Agenda in Ireland." *New Hibernia Review* 6: 138–50.

Braake, T. 2004. "The Dutch 2002 Embryos Act and the Convention on Human Rights and Biomedicine: Some Issues." *European Journal of Health Law* 11:139–51.

Byrne, R., and J. P. McCutcheon 2001. *The Irish Legal System*. 4th edn. Dublin: Butterworths.

Caulfield, T. 2003. "The Regulation of Embryonic Stem Cell Research: A Few Observations on the International Scene." *Health Law Journal* (Special Issue) 87–95.

Council of Europe. 2010. "Steering Committee on Bioethics." Chart of Signatures and Ratifications. Accessed March 8, 2011. http://www.coe.int/t/dg3/healthbioethic/source/INF(2010)1%20%C3%A9tat%20sign%20ratif%20r%C3%A9serves.doc

Cousins, G. et al. 2005. *Public Perception of Biomedical Research: A Survey of the General Population of Ireland*. Ireland: Royal College of Surgeons.

Dickenson, D. 1999. "Cross-Cultural Issues in European Bioethics." *Bioethics* 13: 249–55.

Donnelly, J. 2000. *Realism and International Relations*. Cambridge: Cambridge University Press.

Gunning, J. 1999. "Article 18 of the European Biomedicine Convention: What Impact on Assisted Reproduction?" *European Journal of Health Law* 6: 165–71.

Hervey, T., and J. McHale 2004. *Health Law and the European Union*. Cambridge: Cambridge University Press.

Irish Census 2006. 2007. Volume 3 – Household Composition, Family Units and Fertility.

Irish Council for Bioethics. 2005. "Bioethics Research." Accessed March 8, 2011. http://www.bioethics.ie/uploads/docs/129171-Bioethics%20Research.pdf

Irish Medical Council. 2009. *The Guide to Professional Conduct and Ethics for Registered Medical Practitioners*. 7th edn.

Isasi, R., and B. Knoppers 2006. "Mind the Gap: Policy Approaches to Embryonic Stem Cell and Cloning Research in 50 Countries." *European Journal of Health Law* 13: 9–25.

Madden, D. 2006a. "Assisted Reproduction in the Republic of Ireland – A Legal Quagmire." In *Ethics, Law and Society*, edited by J. Gunning and S. Holm , Vol. 2. Aldershot: Ashgate.

Madden, D. 2006b. "Report of Dr Deirdre Madden on Post Mortem Practice and Procedures." Accessed March 7, 2011. http://www.dohc.ie/publications/madden.html

McDonnell, O., and J. Allison 2006. "From Biopolitics to Bioethics: Church, State, Medicine and Assisted Reproductive Technology in Ireland." *Sociology of Health & Illness* 28: 817–37.

McHale, J. 2005. "The Human Tissue Act 2004: Innovative Legislation – Fundamentally Flawed, or Missed Opportunity?" *The Liverpool Law Review* 26: 169–88.

Mulkay, M. 1997. *The Embryo Research Debate: Science and the Politics of Reproduction.* Cambridge: Cambridge University Press.

Plomer, A. 2002. "Beyond the HFE Act 1990: The Regulation of Stem Cell Research in the UK." *Medical Law Review* 10: 132–64.

Price, D. 2005. "The Human Tissue Act 2004." *The Modern Law Review* 68: 798–821.

Redfern, M. 1999. "The Royal Liverpool Children's Inquiry." Accessed March 8, 2011. http://www.rlcinquiry.org.uk/

Senead, Éireann. Volume 170. 4 December 2002. Adjournment Matters. Oviedo Convention.

The Constitution Review Group. 1996. *Report of the Constitution Review Group.* Dublin: Stationary Office.

The Department of Health and Children. 2009. "Draft Proposals for General Scheme of the Human Tissue Bill 2009." Accessed March 8, 2011. http://www.dohc.ie/consultations/closed/human_tissue_bill/

Walin, L. 2007. "Ambiguity of the Embryo Protection in the Human Rights and Biomedicine Convention: Experience from the Nordic Countries." *European Journal of Health Law* 14: 131–48.

Walters, L. 2004. "Human Embryonic Stem Cell Research: An Intercultural Perspective." *Kennedy Institute of Ethics Journal* 14: 3–38.

Chapter 12
Legal and Ethical Aspects of Biobanks for Research in the European-Mediterranean Area

Renzo Pegoraro, Alessandra Bernardi, and Fabrizio Turoldo

12.1 Introduction

Biobanks are containers of biological samples and associated information that have great relevance for the advancement of research in life science, biotechnology and human health. For instance, population-based biobanks are needed for epidemiological studies and for studies of environmental and genetic factors and human health, disease-oriented biobanks are meaningful in the context of clinical trials and/or multicentric groups with specific objectives and priorities defined upfront. Transfer of biological samples certainly is an inevitable step to generate scientific knowledge.

This chapter will address the questions of procurement, storage, and transfer of tissues and cells for non-clinical research purposes. In particular, we start by describing the current situation of the European Union East-Mediterranean Countries (Cyprus, Greece, Italy, Malta, and Slovenia) and exploring the main similarities and differences between national legislation and ethical guidelines, and the application of EU legislation in each country.

Several ethical issues (e.g. the protection of confidentiality, informed consent) have already found answers in the law, at least partially. In fact, the investigated countries show different levels of development concerning establishment of research biobanks and the related legislation. Slovenia appears the most organised, already having a tailor-made law. Greece and Italy demonstrate an intermediate advancement regarding legislation on biobanks (they do not have tailor-made laws but refer to the national constitution, other not specific acts or guidelines). Maltese legislation is silent on a number of topics relating to research; legislation refers more particularly to the outcomes of research and to research products rather than to the conduct of the research itself. In Cyprus there is no specific legislation relating to biobanks for research but the Cyprus National Bioethics Committee (CyNBC) is playing a key role in the national organization of sample collections and biobanks.

R. Pegoraro (✉)
Fondazione Lanza, Via Dante, 55-35139, Padova, Italy
e-mail: renzo.pegoraro@fondazionelanza.it

C. Lenk et al. (eds.), *Biobanks and Tissue Research*, The International Library of Ethics, Law and Technology 8, DOI 10.1007/978-94-007-1673-5_12,

In the following paragraphs we will examine in detail the provisions regarding the procurement, storage, and transfer of tissue and cells for research purposes in the countries mentioned above. It is necessary to underline that some acts, even if they do not specifically pertain to biobanks for research, will be described in this text. In fact, many directives contained within these laws or guidelines can be useful and are already applied in some biobanks for research purposes.

Subsequently, we will explore the national policies on sensitive data protection and donor's consent. Open issues will be addressed in the last part of this chapter, where we will devote more attention to the ethical dimension of the topic.

12.2 Legal Aspects

12.2.1 General Provisions

12.2.1.1 Slovenia: A Discrete Tissue Law Exists Only in Theory

Slovenia has recently (2007) passed the "Act on Human Tissue". This Act fulfils the requirements and obligations of: the Oviedo Convention of the Council of Europe (1997), Directive 2004/23/ES of European Parliament and the Council of Europe (31 March 2004) on setting standards of quality and safety for the donation, procurement, testing, processing, preservation, storage, and distribution of human tissues and cells, the Commission Directive 2006/17/EC (8 February 2006) which implements Directive 2004/23/EC, and the Commission Directive 2006/86/EC (24 October 2006) implementing Directive 2004/23/EC. The "Act on Human Tissue" has replaced at least fifteen different national directives. This act regulates all treatments, for therapeutic purposes, using tissues and cells of human origin: offering, acquisition, testing, treatment, preservation, allocation, distribution, and the procurement of industrial products on the basis of these tissues or cells. It also regulates all research activities with tissues and cells, including the blood stem cells obtained from medulla, peripheral vein blood and from umbilical cord, from reproductive cells and from adult and embryonic stem cells, intended for therapeutic treatment (Art. 2 of ZKVCTC). Aside from that it should ensure that the persons who perform the law are well qualified. Each tissue, cell, or product has to have a clear origin, traceability, and intention data (Art. 10).

According to the Act all research activities with tissue and cells of human origin, even if anonymous, have to be reviewed and approved before the actual research may be started. Ethical review is also obligatory for research on personal medical data, including epidemiological studies.

The first responsible institution for performing the "Act on Human Tissue" is the Agency for Medicinal Products and Medical Devices (Agency) of the Republic of Slovenia (Javna agencija za zdravila in medicinske pripomočke), which was formed on 1 January 2007 by the merger of Agency for Medicinal Products and Medical Devices (Agencija za zdravila in medicinske pripomočke – ARSZMP).

The term "tissue bank" defines the organisational Unit within the Agency (Art. 17). All organisational Units have to procure full anonymity of all personal data so that the donor as well as the recipient could not be recognised by the third person.

This law should not only prevent the misuse of freedom of research and reduce the lack of organisation in that field but also help to determine the institutional compliance with ethical requirements. However, in Slovenia, a discrete tissue law exists only in theory, not in practice, because the Agency is, as a main responsible institution, totally overloaded and organisationally unadjusted. Slovenian researchers and research institutions are not convinced that the act would be credible enough. General opinion is that the law has overly high expectations towards both the above mentioned Agency and the ethical competency of the researchers and the research institutions involved (Mlinar A: Oral presentation at the Tiss.EU Workshop in Padua, September 2009). Some legal provisions regarding biomedical research are also included in actual Slovenian "Penal Code" (UL RS 95/04) and in the "Act on fertility treatment and procedures of assisted procreation with biomedical assistance" (UL RS 70/2000). A law on protection of human rights with regards to genetic and a law on biomedical research on human being are in progress.

12.2.1.2 Greece: Foundations in the National Constitution

In Greece, with regard to biobanks, there are indications in the national constitution: the protection of dignity and privacy, the protection of health and genetic identity (Art. 5), and the protection of public health are an obligation of the state (Art. 21).

As far as legislation is concerned, the EU Directives 2004/23/EE, 2006/17/EE and 2006/86/EE have been incorporated into Greek law with "Presidential Decree 26/2008" (FEK A'51, 24.3.2008). The Presidential Decree regulates the standards of quality and safety for the donation, procurement, testing, processing, preservation, storage, and distribution of human tissues and cells. However, this Decree does not clearly apply to biobanking as it refers to tissue and cells for medical applications.

Other acts regulating specific aspects of biobanking are: Data Protection Act (2472/1997) (Art. 2: any data related to health is classified as "sensitive data"), Medically Assisted Reproduction Act (3305/2005) (banking of reproductive cells and tissue but for reproductive purposes only), Health System Reform Act (2071/1992) (rights of hospital in-patients), Code of Medical Ethics (Act 3418/2005) (with regards to confidentiality closed and informed consent).

12.2.1.3 Italy: No Specific Acts and Ethical Guidelines

The Italian legislators did not adopt any ad hoc legislation concerning biobanks activities (for exchange of human tissue and cells for research). Italian regulation regarding human samples and associated data is encompassed in data protection acts (laws and decrees) including the exchange (import/export) of biological material for research uses. In particular, the "General Authorisation for the Processing of Genetic Data" issued by the Italian Data Protection Authority (The "Garante per la protezione dei dati personali", later on called the Garante) (22 February 2007). This authorisation applies to the processing of genetic data and also to the collection and use of biological samples for health care and research purposes. Originally, this only applied from April 2007 until 31 December 2008, but it was extended, following the Garante decision of 19 December 2008, to 31 December 2009. It shall be the same for the year 2010.

Another important act is the "Legislative Decree no. 196 of 30 June 2003, Personal data protection Code". This Decree is a "Code" on protection and treatment of personal data in any field, there is a specific section regard the treatment of personal and sensitive data for scientific research or statistical and historical purposes. It provides rules of international flows. The Decree does not mention biological samples and human bodily materials; however, information and data deriving from processing and analysing biological material are clearly to be referred to as sensitive personal health data. This Decree shall not apply to anonymous data.

Besides, a decree by the Ministry of Economic Development establishes the procedure for the institution of organisms delegated to certification of biobanks as biological resource centres (2006).

On the other hand, there are ethical guidelines and opinions from the Italian National Bioethics Committee (CNB) and the National Committee for Biosecurity and Biotechnology adopted on biobanks activities inspired by the European and international regulation (such as the Organisation for Economic Cooperation and Development (OECD) Guidelines on biobanking).

In particular, the "Guidelines for the institution and the certification of biobanks" (National Bioethics Committee, 19 April 2006) contain information on biobanks and give general and technical instructions for the acquisition, maintenance and distribution of biological materials. It also contains information on the control of Biological Resource Centers (BRC) as defined by the OECD. The purpose of these guidelines is to ensure the highest quality and authenticity of the biological material, the safety for research and development in various laboratories and to contribute to health protection of the laboratory staff, the public and the environment. It provides a definition of "biobank" guidelines for good management, certification and criteria to ensure the rights of biological material donors for medical research.

Moreover, the "Guidelines for the recognising and accreditation of research biobanks" (National Committee for Biosecurity, Biotechnology and Life Sciences 2008) provide core information for the accreditation of Italian research biobanks, including storage, distribution, and traceability conditions for biological material. They complement the guidelines of 2006 listed above.

We have also to mention that, in 2003, the document "Genetic Biobanks – Guidelines", issued by the Italian Society of Human Genetics in conjunction with Telethon, have already supplied detailed recommendations on the aims, setting up, management, and the accreditation of biobanks.

12.2.1.4 Malta: Legislation Is Silent on a Number of Topics Relating to Research

In Maltese legislation there is no specific mention of the use of tissues in research (blood, tissues, organs, cells, and DNA). There is, however, a "Human Blood and Transplants Act", and it is envisaged that the EU directive 2004/23/EC be incorporated into this act. The bill has been published as Bill No. 62 in the Government Gazette, No. 17886, 28 February 2006. "This bill incorporates licensing and controls to ensure adherence to the directive, which sets standards of quality and safety for the donation, procurement, testing, processing, preservation, storage and distribution of human tissues and cells, including "haematopoietic peripheral blood, umbilical-cord (blood) and bone marrow stem cells, reproductive cells (eggs, sperm), foetal tissues and cells and adult and embryonic stem cells". The directive only applies to research on human tissues and cells when related to the human body (in vitro research is excluded)" (Cauhci et al. 2006). However, it excludes the use or non-use which the country may adopt for any specific human cells, including those mentioned. It is therefore only a law which sets standards for licensing and safety.

The Bioethics Consultative Committee also issued a document on "Ethical Considerations Relating to Human Reproductive Technology", which made recommendations to the Minister of Health on the use of embryonic tissues for research.

12.2.1.5 Cyprus: The Ethical Point of View Where Law Is Lacking

Cyprus does not have any laws to regulate biobanking. Only the "Law on the protection of personal data" (L. 28(III)/2001) applies where the issue of such a protection arises as far as it regards human biomedical materials and the persons involved.

However, from the ethical point of view, recently the Cyprus National Bioethics Committee (CyNBC) has published its "Opinion on the Establishment and Use of Biobanks and Registries of Human Biological Samples for Research Purposes".

To establish the provisions for the procurement, storage, and transfer of tissues and cells and other biological samples for research purposes the CyNBC felt very strongly that it had to define, first, the term "biobank" and "biobanking" versus "sample collection" or "sample databasing".

Biobanks contain human biological samples and/or substances with or without personal data and other relevant information. Regarding them, CyNBC establishes that the state must legally recognize several terms defined in the following table.

Critical requirements for biobanks – CyNBC Opinion
• The purpose and role of the biobank. • The principles and procedures governing the collection and provision of the samples/data to researchers. • The principles and procedures governing the collection and provision of the samples/data to researchers. • The transparency/dissemination of the research results originating from samples/data of the biobank. • The necessity of free and informed consent from the individuals donating samples/data, who must know that their samples will be placed in the biobank. • The samples have to be collected to support several present and future research projects both within and outside the country. • Researchers from both Cyprus and abroad can have access to the samples/data of the biobank by following established procedures. • The establishment and operation of the biobank should be notified to the Data Protection Commissioner. • A stringent quality assurance procedures (accreditation).

Sample collections or sample databases contain human biological samples and/or substances with or without personal data and other relevant information. Concerning them, CyNBC states that the individual and/or organisation/establishment/institution is responsible, the samples are collected for specific research projects, free and informed consent can be either "closed" or "open", the establishment and operation of the biobank should be notified to the Data Protection Commissioner.

12.2.2 Sensitive Data Protection

12.2.2.1 Slovenia

On 15 July 2004, the Slovenian Parliament passed a new Personal Data Protection Act (ZVOP-1). On 1 January 2005, ZVOP-1 replaced the old Personal Data Protection Act (ZVOP of 1999, as amended in 2001 to render it accordant with Directive 95/46/EC).

This law states that sensitive personal data must, during processing, be specially marked and protected, such that access to them by unauthorised persons is prevented; personal data may only be collected for specific and lawful purposes, and may not be further processed in such a manner that their processing would be counter to these purposes (Art. 16).

Although, irrespective of the initial purpose of the collection, personal data may be further processed for historical, statistical, and scientific-research purposes in an anonymous form, unless otherwise provided by statute or if the individual to whom the personal data relate gave prior written consent for the data to be processed without being made anonymous (Art. 17). The National Supervisory Body for Personal

Data Protection shall have the status of supervisory body for the protection of personal data (Art. 37).

Currently, the Agency and the donor centres in Slovenia are allowed to undertake the procurement, storage and transfer of tissue and cells from biobanks. The Article 10 of the "Act on Human Tissues" provides the conditions for traceability in both directions (from donor to user and back) with strictly considering of personal data protection (necessary double-blind marking of each part of tissue, cells or product).

12.2.2.2 Greece

The 1981 Council of Europe Convention regarding the protection of individuals from the automated processing of personal data was ratified by Greece over a decade after it was signed (Law 2068/1992). However, no specific data protection measures were adopted until 1997. In the meantime, Greek constitutional provisions provided the backbone of the protection of personal data. It was soon recognised that such general provisions were inadequate to deal with the specific complex problems that had emerged. Despite this, successive attempts to introduce a dedicated data protection law in 1985, 1989, 1990, 1991, and 1992 all foundered.

Finally, Law 2472/1997 ("Data Protection Act") was introduced, incorporating Directive 95/46/EC into Greek law, and establishing the Hellenic Data Protection Authority.

Actually, in Greece collection and processing of sensitive data require prior authorisation by the Data Protection Authority. There are exceptions, such as for data collection from physicians that are bound by confidentiality, but these exceptions do not apply to biobanks.

According to the conditions of consent, access can have only named persons. The data can be processed for purposes stated in the information given to the donor prior to his or her consent. Besides, data can be lawfully processed for research and scientific purposes on the condition that it is anonymized. Public Authorities can access records without the subject's consent but following the authorisation of the Data Protection Authority.

The "Data Protection Act" states that storage of data should ensure that it is safe and secure (Art. 10) but it doesn't contain provisions for the storage of biological material. Only the (above mentioned) Presidential Decree regulates the standards of quality and safety for storage of human tissues and cells but, as already said, this decree does not clearly apply to biobanking as it refers to tissue and cells for medical applications.

12.2.2.3 Italy

With regard to Italy, the "General Authorisation for the processing of genetic data" (The Garante per la protezione dei dati personali, 22 February 2007) authorizes the processing of genetic data to natural and legal persons, research bodies and/or institutions, associations, and other public or private bodies seeking research purposes, with regard to such data and operations as are indispensable exclusively for the purposes of scientific research, including statistics, in view of protecting the data subjects, "third parties" and/or the community's health in the medical, biomedical, epidemiological, and anthropological sectors within the framework of the activities falling under the scope of medical genetics (Art. 2.d). The Garante established that such genetic data may be processed as it is particularly relevant to scientific and statistical research with a view to protecting the community's health in the medical, biomedical, and epidemiological sectors, providing that the availability of exclusively anonymous data on population samples does not allow the research purposes to be achieved, whereby said research shall be carried out with the data subject's consent except for the statistical surveys and/or scientific researches provided for by law (Art. 3). Where the purposes for which genetic data are processed may not be achieved without identifying data subjects, also on a temporary basis, the data controller shall take specific measures to keep identification data separate ever since collection – except where this is found to be impossible on account of the peculiarities of the processing or requires an effort that is manifestly disproportionate (Art. 4.1). Any scientific and/or statistical research which makes use of genetic data and biological samples should be carried out in accordance with guidelines established according to the relevant sector-related standards. This should also be done to provide proof that the data being processed and the biological samples being used are for appropriate scientific purposes (Art. 4.2).

12.2.2.4 Malta

The Maltese Data Protection Act of 2001 was the first law in Malta that directs itself exclusively to the protection of personal data. It was introduced in order to render Maltese law compatible with Directive 95/46/EC. The Act came into force on 15 July 2003.

The "Data Protection Act" protects the individual so that any samples obtained be used only for the intended purposes and for which the informed consent was obtained. There is also a time limit set upon which the sample can be stored. Article 8, following the EU Directive, grants exemption when data is processed for historical, statistical or scientific purposes. The term "scientific purposes" remains as unclear as in the EU counterpart. Many would interpret this as meaning that once data is anonymised – and it is guaranteed that individuals cannot in any way be identified by the results – then one may procure, store and use these tissues. Recital 26 of the same Directive makes this doubtful however.

12.2.2.5 Cyprus

"Cyprus National Bioethics Commission Recommendations" establishes that, for biobanks, access should be permitted to any organisation, university, institute or other legal establishment, private or governmental, which obtains approval by the

CyNBC, the permission of the government (Ministry of Health) and notifies the establishment of the biobank to the Data Protection Commissioner.

With regard to sample collections or sample databases, the CyNBC states that access should be permitted to any legal entity which obtains approval by the CyNBC and notifies the establishment of the collection/database to the Data Protection Commissioner.

Referring to the privacy of research subjects, Cyprus assured compliance with Directive 95/46/EC through the implementation of the "Processing of Personal Data (Protection of the Individual) Law" of 2001. The 2001 law was amended in 2003, through Law No. 37(I)/2003.

It is remarkable that in its introduction, the Law defines "processing" or "processing of personal data" as any operation or set of operations which is performed by any person upon personal data, whether or not by automatic means, and includes collection, recording, organisation, preservation, storage, alteration, extraction, use, transmission, and dissemination (Art. 2).

On the basis of this act, the processing of sensitive data, is permitted, when it is performed for statistical, research, scientific, and historical purposes, on condition that all the necessary measures are taken for the protection of the data subjects (Art. 6.1); the controller (any person who determines the purpose and means of the processing of personal data) shall ensure that the personal data are processed fairly and lawfully, collected for specified, explicit, and legitimate purposes and are not further processed in a way incompatible with those purposes; kept in a form which permits identification of data subjects for no longer than is necessary (Art. 4.1); in addition, the processing must be notified to the Data Protection Commissioner (Art. 7.1).

12.2.3 Consent

12.2.3.1 Slovenia

Slovenia does not have specific legal regulations related to informed consent (IC). The "Law on Patient's Rights" (2008) has defined some requirements about information and consenting in medical treatment (full or bidirectional consent, which means that the person involved should receive as well as understand information regarding the procedure of obtaining tissue, as well as about the influence on his/her health/personal condition). In principle, the full IC is required but not in cases where seeking consent would involve unreasonable efforts and if the risk of undue invasion of privacy is minimized (for example researching on archived biological samples if the anonymity would be irreversible). In principle, the use of the body or a part of it for research (scientific purposes) or for pedagogical ends is allowed if the deceased person has explicitly consented before his/her death. According to the actual legislation on removal of parts of the body and transplantation, the use of the body is also allowed if it is known in general that he/she would not be contrary to it. In that case the consent of one relative is needed. If the donor is alive, the full consent is needed. According to the Declaration of Helsinki it is the duty of the physician/researcher to protect life, health, dignity, integrity, right to self-determination, privacy, and confidentiality of personal information of research subjects. All these elements are also parts of full consent so that the participation of competent individuals as subjects is voluntary.

12.2.3.2 Greece

In Greece, collection of biological material without consent is unconstitutional. The Data Protection Act requires that consent is "freely given, explicit and specific indication of will". Data subjects should receive "information as to the purpose of processing", "the data or data categories being processed", "the recipient or categories of recipients of personal data as well as the name, trade name and address of the Controller and his/her representative, if any"; besides, such consent may be revoked at any time without retroactive effect.

12.2.3.3 Italy

In Italy, written informed consent to export biological material and associated data for research purposes is required from the sample donor (Legislative Decree 196/2003, Section 43 and 77, Guidelines on biobanks 2006, point 3.7; Guidelines of 2009 on informed consent).

Human biological material and genetic information associated must be treated like any other medical information, as "personal sensitive data" and therefore it is forbidden to give this information to third parties, another European country or third country, without written explicit consent from the biological material donor.

General donor's informed consent in Italy – basic information required

- Research purposes, storage duration, uses of biological sample
- Localisation of the biological sample storage and any shipments of samples (here is the export specification) within the scope of the study and/or of the biobanking project
- Risks and benefits for the donor and/or the society/institution
- Identification of all entities involved in data processing with their roles and responsibilities
- The "Titolare" of the personal data treatments and the "Responsibles" of the Institution and/or biobank, their addresses and telephones
- Assurance that the purpose of data processing is fair, lawful and legitimate
- Implementation of precautionary security measures to reduce risks on data disclosure
- Biological material and sensitive data accessibility
- Commercial uses of the output (if there is)
- Results of the research
- Possibilities of withdrawal consent (on the storage, the use or the transfer of samples and data) and the right of the donor to oppose to the use for legitimate reasons

- Explicit authorisation by the donors to the "treatments" of his/her personal data on the basis of the Legislative Decree n. 196/2003 (consent without this legal specification is not valid).

The Italian National Bioethics Committee together with the National Committee for Biosafety and Biotechnology provide a model of consent (National Bioethics Committee and National Committee for Biosafety and Biotechnology Recommendations on informed consent concerning human biological materials used in research purpose 2009).

Consent is not required when the processing concerns completely anonymous data (Legislative Decree 196/2003).

Informed consent for secondary use of human biological material already stored shall be required in principle when a new use (as export and external storage and uses) of the human biological sample and data was not originally planned on the first consent (National Bioethics Committee Guidelines for the certification of Biobank of 2006, point 3.9 and Guidelines of 2009 on informed consent). The donor's informed consent is required when data are directly or indirectly identifiable; however, when tissues and cells contain anonymous data the material could be used in secondary medical research without seeking the consent of the donors in retrospect.

In case of doubt on the necessity of a new informed consent for a secondary use, the Institution carrying out the research or managing the biobank has to seek the Ethics Committee's advice. When scientific data have been produced with the consent of a person, the same person should not have the right to ask for their destruction but only the anonymization, though she/he may always ask for the destruction of the biological samples.

12.2.3.4 Malta

The issue of informed consent in research is not dealt with in Maltese legislation. Malta is expected to sign and ratify the Council of Europe Convention on Human Rights and Biomedicine in the near future, following which the relevant articles relating to research will apply to Malta.

As mentioned above, the Maltese Data Protection Act protects the individual so that any samples obtained be used only for the intended purposes and for which the informed consent was obtained. The Data Protection Act defines "consent" as "any freely given specific and informed indication of the wishes of the data subject by which he signifies his agreement to personal data relating to him being processed". This does not refer specifically to research, but to any collection of data from a data subject. This also includes processing operations in cases of research.

12.2.3.5 Cyprus

The "Cypriot Processing of Personal Data (protection of individuals) Law 138 (I) 2001" (amended in the Law 37 (I) 2003) establishes that "personal data may be processed only if the data subject has unambiguously given his consent" where "consent" means consent of the data subject, any freely given, express, and specific indication of his wishes, clearly expressed and informed, by which the data subject, having been previously informed, consents to the processing of personal data concerning him.

Donor's informed consent – CyNBC Opinion

- Donor has to be informed that his/her sample/data would be stored for research purposes (with "open" or "closed" consent) and that his/her sample/data may end-up in different research teams in different parts of the world.
- Donor has to be given the opportunity to declare whether he/she would like to be informed of any research result that may directly or indirectly affect his/her health.
- Donor has to be told what would happen to his/her sample/data in case the biobank closes down.
- Donor has to be informed that his/her has the right, at any given time, to withdraw his/her sample/data.
- The consent form is approved by the CyNBC or its Ethics Review Committees.

Regarding the secondary use of samples (i.e. research endeavors diverging from the original foreseen purpose), for all samples collected after 2004 (operation of the CyNBC), all researchers must ask the donors of the samples to declare whether they provide either a "closed" (for specific project with defined duration) or an "open" (when samples/data can be used also for other future research programs provided these are approved by the CyNBC and its established Ethics Review Committee) consent. Samples collected, legally, before 2004 can be used for research provided the research activity is approved by the CyNBC and the samples are anonymized.

12.3 Ethical Aspects

12.3.1 Ethical Values

What are the most relevant ethical values involved in biobanking? We mainly focused on the values of *gift* and *trust*. First we asked ourselves: what is the real meaning of "gift" and what are the implications of this particular meaning with the gift of a tissue to a biobank? Our first answer was that the donation of a gift is something different from selling something. When someone sells something, he loses forever every kind of relationship with the object and with his new owner. If you sell a ring to a woman, then it's not your business what this woman does with the ring: she could destroy it, or sell it to someone else, or even use it as a gift for another person. With the act of selling, in fact, you lose forever every emotional connection with the ring. But, if you give a ring as a present to a woman, you certainly lose your right of property over the ring, but, at the same time, you don't lose every emotional connection with the object. The woman can do whatever she wants with the ring, but you can complain if she destroys it, or puts it into the sewer, because this ring is not your property, but the relation of gift gives to the ring something special: an emotional connection with you.

We think that something similar could happen with the donation of tissues: if these tissues are the result of a gift, biobanks have the moral duty to use them in a proper way and, at the same time, they have the duty to be extremely transparent before public opinion.

Another important value at stake, which is linked with the other one mentioned before, is trust. Sociologists involved in our research group gave us evidence that people aren't inclined to give their tissues to every kind of biobank and to support every kind of research. It depends on public opinion and trust in some particular areas of research, and on the moral authority of the scientists involved. Therefore we think that a lot of issues involved in biobanking as, for example, respect of privacy, informed consent, etc., are important means aimed to safeguard a more radical and basic value, which is trust.

12.3.2 Open Questions: Between Law and Ethics

Can legislative acts resolve all ethical issues on biobanking? If not, what issues remain unsolved? Should you define all by Law or some questions can be better managed only at the ethical level?

From the analysis of legislative acts we can discover points of convergence and diversity within each country.

Genetic data are generally considered "sensitive data" and their protection has several common provisions in many acts describes above, mainly because, as established by the European Union in order to harmonise data protection regulation, member states have transposed the Directive 95/46/EC into internal laws.

Consent remains one of the main issues when considering legal and ethical regulations on biobanks, because consent entails two interests which are legitimate and opposite at the same time: the right of the source subject to protect his/her personal data and the interests of the scientific community. The development of biobanks in the East-Mediterranean area has created controversy, particularly regarding the importance and procedures of informed consent. The consent of participants is usually required before biobank samples can be used in research, but the nature of this consent, and how and when it is obtained, varies widely.

Slovenian law establishes that full informed consent is required but not in case if seeking consent would involve unreasonable efforts and if the risk of undue invasion of privacy is minimized (for example researching on archived biological samples if the anonymity would be irreversible).

Cyprus takes the view that general consent is acceptable to use samples for future, as yet unspecified, research projects on condition that the research activity is approved by Research Ethics Committee and the samples are anonymized. On the same wavelength, Greece states that it would not be unthinkable to present donors with an option between a specific or a "blanket consent" (it covers any use of the material at any time in the future) provided, in the latter case, that their data will be anonymized or encrypted.

In Maltese regulation, any samples obtained have to be used only for the intended purposes and for which the informed consent was obtained.

In Italy, consent is not required when the processing concerns complete anonymous data but informed consent for secondary use of human biological material already stored shall be required in principle when a new use of the human biological sample and data was not originally planned on the first consent.

On the whole, a major ethical problem for prospective biobanks remains how to assure participants' consent, autonomy and dignity when it is not known what they are consenting to in terms of future research.

Besides, other aspects are still debated:

1. the relationship between freedom and interest of researchers and institutions on biobanking and the safeguard of citizens and community's rights;
2. training and responsibility of biobanks' administrators;
3. the role and responsibility of Research Ethics Committees in the evaluation of researches on human biological samples.

12.4 Conclusions

In the examined country group, the application of EU legislation (in particular the EU Tissue Directive and the EU Data Protection Directive) creates a good uniformity on many points. Although, we should consider that the EU Directives and consequently the local laws do not refer specifically to research biobanks. Besides, several aspects remain distinguishing of local choices. Some countries define many issues by law (Slovenia) or seem relatively bound by EU legislation (Malta); in other cases, the ethical approach is predominant (Cyprus) or the country refers to ethical guidelines as to laws (Greece, Italy).

In our opinion, decisive answers on some issues should not be given by law. In fact, enforceable standards cannot be applied to ethical questions that require continuous debate and flexible solutions. We recognise, for instance, that the protection of privacy interests must be balanced with the interest in advancing research, even if it is rather complicated. This goal can be achieved in different ways: defining the informed consent model (e.g. a specific consent, a partially restricted consent or a multi-layered ones instead of a "blanket consent"), increasing public participation and trust, establishing ethical and scientific supervision of biobanks and research projects, giving different rules for different types of data and projects. It seems desirable to have a European reflection on this topic and on other sensitive themes related to biobanks. We underline the Cypriot approach which appears very interesting despite the limits connected with the size of this country, in particular because it demonstrates the possibility and usefulness to use ethical guidelines instead of law.

References

Slovenia

Act on Quality and Safety of human tissues and cells intended for medical treatment (Zakon o kakovosti in varnosti človeških tkiv in celic, namenjenih za zdravljenje – ZKVCTC) (Act on Human Tissues) (UL RS 61/07).
On Removal and Transplantation of parts of Human Body for Medical Treatment Act (Zakon o odvzemu in presaditvi delov èloveskega telesa zaradi zdravljenja). UL RS 12/00.
On Fertility Treatment and on Procedures of Assisted Procreation with Biomedical Assistance Act (Zakon o zdravljenju neplodnosti in postopkih oploditve z biomedicinsko pomoèjo). UL RS 70/00.
Personal Data Protection Act (Zakon o varovanju osebnih podatkov) UL RS 86/04.
Penal Code of RS (Kazenski zakonik Republike Slovenije). UL RS 95/04.
On Quality and Safety of Human Tissues and Cells Intended for Medical Treatment Act (Zakon o kakovosti in varnosti èloveskih tkiv in celic, namenjenih za zdravljenje). UL RS 61/07.
On Patient's Rights Act (Zakon o pacientovih pravicah). UL RS 15/08.

Greece

Law 2472/1997 on the Protection of individuals with regard to the Processing of Personal Data (as amended). Available at: http://www.dpa.gr/portal/page?_pageid=33,43560&_dad=portal&_schema=PORTAL (accessed 10 March 2011).
Act 3418/2005 (Code of Medical Ethics).
Act 3305/2005 (Medically Assisted Reproduction).
Act 2071/1992 (Health System Reform).
National Bioethics Commission (2002). Opinion on the collection and use of genetic data. Available at: http://www.bioethics.gr/ (accessed 10 March 2011).
National Bioethics Commission (2006). Opinion on banks of biological material of human origin (biobanks) in biomedical research. Available at: http://www.bioethics.gr/ (accessed 10 March 2011).
National Bioethics Commission (2007). Opinion on umbilical cord blood banking. Available at: http://www.bioethics.gr/ (accessed 10 March 2011).

Italy

The Garante per la protezione dei dati personali (2007). General Authorisation for the Processing of Genetic Data – 22 Februray 2007 (Autorizzazione al trattamento dei dati genetici – 22 Febbraio 2007) Gazzetta Ufficiale (Italian Gazzette) n. 65 dated 19 March 2007. Available at: http://www.garanteprivacy.it/garante/doc.jsp?ID=1395420 (accessed 10 March 2011).
Legislative Decree n. 196 dated 30 June 2003 (Personal data protection Code). Available at: http://www.garanteprivacy.it/garante/navig/jsp/index.jsp (accessed 10 March 2011).
Comitato Nazionale per la Biosicurezza e le Biotecnologie (National Committee for Biosecurity and Biotechnology) (2006) Linee guida per la certificazione delle biobanche. Available at: http://www.governo.it/biotecnologie/documenti.html (accessed 10 March 2011).
Italian Society of Human Genetics and Telethon Foundation (2003) Biobanche genetiche linee guida (Genetic Biobanks Guidelines). Available at: http://www.sigu.net/index.php?option=com_docman&task=cat_view&gid=46&Itemid=78&limitstart=10 (accessed 10 March 2011).

Italian National Bioethics Committee (2006) Biobanks and research on human biological material. Available at: http://www.governo.it/bioetica/eng/opinions.html (accessed 10 March 2011).

Comitato Nazionale per la Biosicurezza, le Biotecnologie e le Scienze della Vita (National Committee for Biosecurity, Biotechnology and Life Sciences) (2008) Biobanche e Centri per le risorse biologiche di campioni umani, istituiti a fini di ricerca: linee guida per il riconoscimento e l'accreditamento (Guidelines for the recognising and accreditation of research biobanks). Available at: http://www.governo.it/biotecnologie/documenti/linee_guida_definitivo_2008.pdf (accessed 10 March 2011).

Malta

Cauhci, M. C., Aquilina, K., and Ellul, B. (2006) *Health, Ethics and the Law*. Malta: Malta University Press.

Human Blood and Transplants Act (Act IV of 2006). Available at: http://www.sahha.gov.mt/pages.aspx?page=974 (accessed 10 March 2011).

Data Protection Act (2001). Available at: http://docs.justice.gov.mt/lom/legislation/english/leg/vol_13/chapt440.pdf (accessed 10 March 2011).

Maltese Bioethics Consultative Committee (2000). Cauhci, M. C., Mallia, P., Agius, E., and German, L. J. (Eds.). Ethical Considerations Relating to Human Reproductive Technology, in Patients' Rights, Reproductive Technology, Transplantation, Symposium Proceedings.

Cyprus

Cyprus National Bioethics Committee (2009).Opinion on the creation and use of biobanks and human origin biosamples archives for research purposes. Available at: http://www.bioethics.gov.cy/Law/cnbc/cnbc.nsf/All/5FE2C627FE9884A1C225772200283F8E?OpenDocument (accessed 10 March 2011).

The Law 138 (I) 2001 (amended in the Law 37 (I) 2003) on Personal Data Processing (Protection of the Individual). Available at: http://www.dataprotection.gov.cy/dataprotection/dataprotection.nsf/index_en/index_en?opendocument (accessed 10 March 2011).

Chapter 13
The Circulation of Human Body Parts and Products: When Exclusive Property Rights Mask the Issue of Access

Florence Bellivier and Christine Noiville

Fifteen years ago, Jean-Pierre Baud, a French legal historian, told us a story to help us understand the complexity of the legal framework of the human body. This piece of legal fiction described the case of the stolen hand (the title of his book): a Sunday home handyman accidentally cut off his hand; he passed out, and an enemy seeking vengeance seized the hand and threw it into the building boiler.[1] Can we consider that he stole it? To answer this question, the author goes across time to Roman law and its masking of the body, which it replaced with the person, the subject of abstract law. He goes on to review the status of relics in the Middle Ages and, at the end of a brilliant demonstration, concludes that the only framework of rules applicable to the human body is that of objects or things, even though this unique and original thing must be considered holy.

With the distance time has provided, this demonstration appears as masterly as it is incomplete:

- *Masterly*, because although the French legislature did not specifically ratify this analysis, the legislative subtext, practices, and doctrine make it irrefutable (I).
- *Incomplete* in relation to trends in the same practices, for they appear to make an analysis in terms of property rights difficult or even misleading, by obscuring the real issue of the use of human body parts and products, i.e., access (II).

13.1 The Appeal of Property

History has undoubtedly proved the qualification of the body as a sacred thing and, hence, the recognition of commodification of body parts, even though the legislature did not confirm this analysis in so many words in Law no° 94-653 dated

Translation by Jo Ann Cahn.

[1] Baud (1993, passim).

F. Bellivier (✉)
University Paris Ouest Nanterre la Défense, Paris, France
e-mail: florence.bellivier@wanadoo.fr

C. Lenk et al. (eds.), *Biobanks and Tissue Research*, The International Library of Ethics, Law and Technology 8, DOI 10.1007/978-94-007-1673-5_13,

29 July 1994. On the contrary, it placed the human body in the Civil Code chapter on persons (art. 16 et seq. Civil Code), it stated in article 16-1 that every person has the right to have his or her body respected (rather than the right to dispose of it freely), and it replaced the principle that the body is inviolable and beyond commerce, which might have been thought to reflect on positive law before 1994, by a principle of non-patrimoniality (art. 16-1 al. 3 and art. 16-5 of the Civil Code), which simply forbids individuals from receiving money directly for the gift of their body, its parts or products. Recognizing the existence of practices that it sought to authorize under strict regulations (donations of human body parts and products, assisted reproductive technology, etc.), the legislature did not want to make the body an object subject to pure and simple alienation, for it remains at the service of the person entitled to it by law (art. 16: "The law ensures the primacy of the person"), nor did it want to entirely ban practices that might be considered to damage the integrity of the human body or objectify it. It therefore set up a system of limited objectification – limited by the dual need of the subject's consent and medical justification for the damage (art. 16-3 Civil Code). The major consequence of this system, however, is commodification, albeit limited. Although an attribute of the person, the body, has clearly been placed in what we might call "legal commerce", defined as intercourse in the affairs of life (that is, all of the exchanges possible between people, with or without consideration, gifts and sales, etc.). The fact that these exchanges are regulated by law does not change that fact. French law has confirmed that the body is a reservoir of material useful for the development of practices that demand oversight. To the human body, tangible and visible manifestation of the person who is the subject of the law, is thus added the human body, object of diverse practices, either as a whole and in its entirety, or in parts (body parts and products). If neither the human being as such nor even (with several specific exceptions) his or her entire body are in legal commerce, its parts and products most assuredly are. Insofar as it is assigned to the person, the human body follows the legal regime (rules) of the person, but when it is detached for one reason or another, it is assigned to the framework of things. Not just any thing, but protected things, of the kind the law knows so well – graves, family memories, public property, cultural heritage, etc. But the legislature has adopted a pragmatic approach, preferring to regulate practices rather than ignore them. It is for this reason that the broad principles affirmed in the Civil Code are described in greater detail in the Public Health Code which is unambiguous on this point: even as it monitors and controls the practices surrounding the circulation of human parts and products, it talks about their distribution and their transfer. The groping, hesitant regulation that characterizes this control is simply the consequence of making human parts and products into objects, if not commodities. Any decree or decision in this field should convince the reader: setting prices for the transfer of blood, of sperm or cardiac valves, approval of good practice guidelines for the transformation and transfer of tissue of human origin, organization of the importing and exporting of organs or of cells, etc. At the heart of the standards related to human body parts and products, the key words today are transfer, distribution, importation, service, and exchange.

European Community law furnishes additional evidence of what J.-P. Baud accurately foresaw as the introduction of human life into the market. "As a matter of principle," as the Council says, neither the human body nor any of its components are goods or merchandise as defined in European Community law, that is, "products that can be assessed in money or likely, as such, to be the object of commercial transactions."[2] In practice nonetheless, it is precisely as such objects of commercial transactions that cells, tissues, blood, gametes and other products move throughout the Community market. They do so on two bases.[3]

First, parts and products of the human body circulate as support for the provision of services, under the principle of the freedom to provide services enshrined in Article 39 of the EC Treaty: a standard medical contract or the provision of assisted reproduction technologies are thus services for the successful completion of which cells and embryos can circulate, and the materials are thus accessories to the services in question.[4]

Secondly, an ever growing number of legal rules derived from these rights organize the conditions in which these human parts and products circulate. One example is Directive 2004/23/EC, intended to facilitate the movement of tissues from one Member State to another; it sets up a system of mutual recognition to limit restrictions of exchanges in this domain.[5] Yet another, still older, example is provided by human blood and plasma. Since 1984, they have been "raw materials" and their derivatives "drugs". As such, they are part of the distribution channels of the pharmaceutical industry and are governed by the laws of the market.[6]

European Community case law certainly did not invalidate this analysis when it decided that the use of a kidney improperly prepared for transplantation falls within the scope of Council Directive 85/374/EEC of 25 July 1985 on the approximation of the laws, regulations and administrative provisions of the Member States concerning liability for defective products.[7] Although we may well continue to shilly-shally about whether body parts and products fall under the classification of "goods" under EU law, we cannot deny that this law governs their movement, including, sometimes, for profit.

---

[2]ECJ, *Commission v. Italy*, 10 December 1968, C-7/68.

[3]Bergé (2002) .

[4]See the analysis of the English court in the case of Diane Blood (Court of Appeals, R v Human Fertilisation and Embryology Authority, *ex parte Blood* [1997] 2 All ER 687), commented by J.-S. Bergé, "Droit communautaire, biomédecine et biotechnologies: entre concordance et antinomie", cited above. See also Sefton-Green (2000), Hunter-Henin (2001, 112).

[5]Directive 2004/23/EC of the European Parliament and Counsel, of 31 March 2004, setting standards of quality and safety for the donation, procurement, testing, processing, preservation, storage and distribution of human tissues and cells, *OJEU* L 102 of 7 April 2004, p. 48 et seq.

[6]Council Directive 89/381/EEC of 14 June 1989 extending the scope of Directives 65/65/EEC and 75/319/EEC on the approximation of provisions laid down by law, regulation or administrative action relating to proprietary medicinal products and laying down special provisions for medicinal products derived from human blood or human plasma, *OJEU* L. 181, 28 June 1989, p. 44.

[7]ECJ, *Veedfald*, 10 May 2001, C-203/99, *Dictionnaire permanent bioéthique et biotechnologies,* 10 September 2001, bulletin 105, p. 7435, *D.* 2001, p. 3065, note P. KAYSER.

To take into account the practices as well as the law on this issue of humans' ownership or property rights over their own bodies, is it necessary to go so far as to say that the human body as a whole is an object and to recognize the subject's ownership of this object? Doctrine is divided on this ancient question, more philosophic than legal, of humans' ownership or property rights over their own body.

- Some stress the distance that ownership sets up between the subject and its object and refuse to go that far. They argue that the "parts of the human body (and the person's legal status) being ab initio indissolubly linked to the subject of the law, to the 'person'; they cannot be considered as exterior and therefore cannot be attached by the effect of property law. Individuals, as subjects, are quite sufficient to ensure their own protection and to exclude others from all that is considered an indissociable component part' of themselves.[8] According to this concept, the person is protected as a person, and the body is the concrete expression, the embodiment, so to speak, of the person. It is thus futile, even redundant, to recognize a property relationship between the self and the body. Thus, it is simply because the law recognizes that people are inviolable and inalienable that they cannot be sold. At the very most, the partisans of this position concede, as lip service only, that human body parts and products removed from the body "can no longer be considered purely and simply as indissociable parts of the person" and that "throughout the entire time that they are separated from a human body, only the concepts of property and opposability *erga omnes* (in relation to all) make it possible to attach them legally to a subject of law, to whom they belong as his or her own, to the exclusion of all others, and to grant them accordingly effective protection against the activity or actions of others (a more or less complete restriction on the right to dispose of something is not incompatible with the existence of a property right or ownership)."[9] Thus, only the property right that I might have over the tumor removed from my body, a right derived from the need to protect the person, always preeminent in the legal system, could explain that I am entitled to know what will be done to my tumor or to demand its destruction.
- Others, relying on a particular conception of property, not as a real right but as a relation of belonging that allows the subject exclusive rights over the object,

[8] "les éléments du corps humain (et les droits de la personnalité) sont (étant) *ab initio* indéfectiblement liés au sujet de droit, à sa 'personne', ils ne sont pas considérés comme extérieurs et n'ont donc pas à lui être rattachés par l'effet du droit de propriété. La personne, en tant que sujet, se suffit à elle-même pour assurer sa propre protection et exclure autrui de tout ce qui en est considéré comme un élément constitutif et indissociable". Danos (2007, and the numerous references in the page notes).

[9] "ne peuvent plus être considérés purement et simplement comme des éléments indissociables de la personne 'et que' pendant toute la durée où ils sont séparés d'un corps humain, seules les notions de propriété et d'opposabilité *erga omnes* permettent de les rattacher juridiquement à un sujet de droit -à qui ils appartiennent en propre à l'exclusion de tout autre- et de leur conférer ainsi une protection efficace contre l'activité ou l'action d'autrui (la restriction plus ou moins complète du droit de disposer n'étant pas incompatible avec l'existence du droit de propriété)". Ibid., p. 229 and pp. 229–30, n° 207.

that is, to exclude others from its enjoyment and disposition, are thus comfortable labelling this relation as one of property and ownership. They do not hesitate to analyze articles 16 et seq of the Civil Code about the status of the human body as treating the body as the object of a property right: "ownership exists and continues as long as a person has the power in principle to accede to all the uses of the object. It does not matter that the law forbids access to a particular use: the owner keeps title from the moment that he has the power to accede in principle to all the other uses. This schema is exactly the one that applies to the human body".[10] According to this analysis, it is not important whether or not all of the characteristics of ownership are united in the hands of the owner, from the moment that in principle he or she uses the thing in question, which is the case for the person and the parts and products of the body. Armed with this framework, these authors thus analyze with perfect relevance all of the practices to which the human body, as a whole or in pieces, is subject, and they succeed without artifice in fitting them into the prism of property law.[11] A single example suffices to demonstrate the accuracy of this analysis. On the subject of the right to bodily integrity, they write, "What is essential in the right to bodily integrity is the monopoly that the law recognizes in the subject's will in this domain. The human body is an object reserved to the person, who sovereignly defines the use to be made of it and the access that others may have. Consent must be understood not in the sense of contract law but rather of property law: It is the manifestation of the exercise of the sovereign power that the subject has over his body."[12] This point of view, we note, would bring French and Anglo-American law closer together, since the "bundle of rights" of the latter makes it possible to recognize that these different uses of and rights over the same item may be divided between several holders or, on the contrary, held together.

In France, the legislature has carefully refrained from choosing between these different concepts of the relation between the person and the body (property or not). Jurisprudence is rare on this topic in France, but has shown itself rather hostile to the concept of property applied to the human body, as we see from a decision of the Administrative Court of Appeals of Douai, 6 December 2005 (M. et Mme T., n° 04DA00376, D. 2006.Pan.1205, obs. Galloux). In this case, the couple had undergone in vitro fertilization. The protocol used called for the storage of several surplus

[10]"un droit de propriété existe et perdure dès lors qu'une personne est investie du pouvoir de principe d'accéder à toutes les utilités d'une chose; peu important que la loi interdise l'accès à telle utilité particulière: le propriétaire conserve son titre du moment qu'il est investi du pouvoir d'accéder par principe à toutes les autres utilités. Ce schéma se retrouve exactement s'agissant du régime du corps humain". Zenati-Castaing and Revet (2006, 240).

[11]Ibid., p. 239 et seq., n° 277 et seq.

[12]"ce qui est essentiel dans le droit à l'intégrité corporelle, c'est le monopole que la loi reconnaît à la volonté du sujet dans ce domaine. Le corps humain est un objet réservé à la personne qui définit souverainement l'usage qui en est fait et l'accès que peut y avoir autrui. Le consentement doit se comprendre non au sens du droit des contrats mais à celui du droit de propriété: il manifeste l'exercice du pouvoir souverain dont le sujet est investi sur son corps". Ibid., p. 244, n° 281.

embryos pending implantation. A technical problem (a crack in the cylinder) led the medical team to inform the couple that it would be better not to use those fertilized eggs, which might have been damaged by the fluid loss that resulted in increasing the temperature in the container and thus, according to current scientific knowledge, harming the nine embryos. The couple then instituted proceedings against the hospital to establish its administrative liability.

The trial court denied their claims, on several bases. What matters here is that the judges refused any compensation for material damage, because the human body and its parts could not be the object of a property right (C. civ., art. 16-1). These judges nonetheless awarded them compensation, set at €10,000, for "*various problems in their living conditions in connection with this incident*." The Court of Appeals was still more severe. Expressing its opinion on the question of "problems in their living conditions" claimed by the couple following the destruction of the frozen embryos, the Court denied that these fertilized oocytes had any patrimonial value. It held that these oocytes could not be the source of a compensable injury. On appeal, therefore, the couple lost any right to any compensation.

The contrast with the approach taken by English judges in a similar case, Yearworth v N Bristol NHS CA 2009,[13] is striking. In that case, six men being treated for cancer at a Bristol hospital had sperm frozen there for future use, before beginning chemotherapy that might impair their fertility. All six signed the required consent forms. Between 28 and 29 June 2003, before the sperm straws could be used, the liquid nitrogen level in the storage tanks fell below the required level, thawing the sperm and making them unusable. The six plaintiffs were not all in exactly the same situation: some regained their fertility naturally, another died, some claimed mental suffering, others a psychiatric disorder.

Regardless of these disparities and of the men's actual possibility of having biological children, the case before the Court was based on tort claims (of negligence) for which the claimants sought damages. The Trust defended itself, not by denying negligence on its part (the automatic system for topping off the nitrogen level was not operational, and no manual procedure was applied to replace it), but by claiming it could not be considered liable: it argued that no psychiatric harm or mental suffering was proved, nor any causal association with the negligence. Most especially, however – and this was the focus of the Court's discussion – according to the Trust's attorneys, the claimants' damages were not compensable by a negligence action for personal injury.

Overruling the trial judge, the Court of Appeals found for the claimants, relying on the claimants' ownership of their sperm: first, they had generated and produced this substance from and by their body, and second, they had arranged to have it stored so that they could use it in the future, even though this usage is highly controlled by technical and administrative regulations that apply to gametes and that

[13] Available at: http://www.bailii.org/ew/cases/EWCA/Civ/2009/37.html (accessed 10 March 2011).

place the medical profession at the center of the system. As proof of this right of control, the Court pointed to the depositor's right to have the sperm destroyed. Finally, the claimants were the only holders of any rights to the sperm, even though these rights might conflict with the duties of the license-holders, for example, the duty to destroy it at the end of a given period of time.

We thus see here that the property-based conception of the relation between the person and their own body may be gaining ground because it is convincing in its entirety, accurate in its understanding of the legal relations that arise in relation to the human body as a whole (organ donations, suicide, euthanasia, etc.) or at the moment of the donation or use of body parts and products (sperm donation, identification of individuals by their genetic profiles).

13.2 Property Versus Access

However appealing the ownership concept of a person's relation to her own body may be, today it appears incomplete in regard to what probably constitutes the major issue in the analysis of the human body: conceptualizing what happens, not at the beginning of the chain of the use of the parts and products of the human body, but following the changes and exchanges to which these human body parts and products give rise. Beyond the theoretical question of self-ownership, beyond the power to exclude others from these parts and products, it is the use of the resource and the sharing of the resulting benefits that give meaning to – and a reason to reconsider – the question of ownership. The traditional questions are no longer relevant, disconnected as they seem today from the real issues: what is the point of recognizing that humans own their body or parts of it? It is further along the chain that we return to the issue of who holds the rights derived from the use of a particular resource.

To illuminate the difficulties of the question, let us extend Baud's story of the stolen hand. Imagine that the vengeance-seeker does not throw the hand in the boiler, that he is a surgeon, and that he returns remorsefully to the site of the accident, has the wounded man transferred rapidly to a hospital, and tries to graft the hand back on the arm. He succeeds. He nonetheless removes the ring finger after discovering a suspicious beauty mark on it during the surgery. What should be done to this finger? Should it be destroyed as surgical waste or saved in a jar? Should the presumed tumor be removed and deposited it in the hospital biobank, where it will be cut into thin slices, analyzed biologically and genetically, and then stored? Does the surgeon need the patient's authorization for these procedures? And if researchers from another hospital are conducting research on melanoma, can they have access to the tumor of this poor home handyman? Under what conditions? Is a new consent required, if indeed there had been a first consent? Does the biobank have the right to make the samples available to the researcher exclusively for only a limited time? How can the patient's health data be kept confidential? If the police investigate this suspicious case and want access to the ring finger, can they have it? If the researcher who is using the finger makes an astounding discovery about the genetic origin of melanoma, develops a diagnostic kit, and obtains a patent for the test, do the various participants in this process – the patient, the surgeon, and the biobank – have any rights to the royalties that the inventor receives?

These are all questions to which this article provides no answers but which are decisive for understanding the trends in medical practices today and the need for their regulation. Although it is true that ownership is fundamentally the power to exclude others from using the object, it is not always the appropriate paradigm for conceptualizing these difficulties, for two principal reasons.

In the first place, although it is eminently relevant for determining if someone can make the resources of his or her body available or, on the contrary, remove them from circulation, in other words, for bearing in mind that at the origin of the circulation of human body parts and products, there is always a subject, and that he or she remains sovereign, property law fails to consider an essential circumstance: most of the time, once the body parts and products have been exchanged, individuals no longer have any interest in reserving use to themselves. What they want, in principle, is for this resource to be used, in their own interest (to treat their disease, for example) or for the interest of the community, understood more or less widely. The famous Moore case, now more than 20 years old, seems rather outdated in this respect, at least in its issues: if the patient had claimed a right of ownership over parts of his body being lucratively exploited by the physician and medical center, it was that they had been taken without his knowledge, a practice that no longer exists, at least in the many states that require informed consent before removing or sampling any body part. Today, what matters to patients is not so much their right to oppose this type of practice at any price as their right to know what will happen to their body parts and, more precisely, who will profit from them: will they be used for research, will that research lead to a new drug, who can benefit from the drug, and under what conditions? The individualist dimension of the paradigm of exclusive property rights thus makes it inadequate.

In the second place, the question and the limits of exclusivity emerge again, but downstream from the subject. Imagine, for example, two researchers who want access at the same time to the resource; or still later on, several subjects who demand their share of the potential benefits from the resource. As legitimate as their respective claims to a title to this or that right may appear, the right is no longer anchored in property law or ownership: no one would claim that a researcher's right of access to a given biological sample originates in the ownership rights he has in it, because he has no original right, such as that a patient might have over something removed from his body, flowing from some pseudo-property right that as the owner of his body. The researcher's right is a disembodied right, in the sense that no previous (biological) foundation authorizes its recognition, such as the relation that individuals have with their bodies. Its legitimacy comes not from a biological basis, such as the innate belonging of the self to the body, but from the work he has done on the resource, a basis that is less immediate and more questionable.

Far from a model of sovereignty, we approach instead another paradigm, that of access. It is not the classic sense of ownership, defined as the right to have access to all the uses of the object and to forbid others any access. The meaning applies rather to a resource that is considered vital or essential and more or less rare, in quantity or quality – information, water, genetic resources, or drugs, for example – and to which several people – the producer of the resource, its user, etc. – can claim materially and legally to have access. Jurists, following the economists, who were first to identify the specificities of the "age of access",[14] thus have the arduous task of organizing these rival rights to access. It is precisely for that reason that they must keep their distance from concepts of ownership and property rights, for these fail to provide an adequate answer. Because they are inevitably attached to an owner, a holder, they finish by blocking the uses of the human body parts and products, putting them into the hands of a protagonist. The issue, on the contrary, is to organize access to it, to ensure the smoothest possible circulation, even though that requires a simultaneous analysis of how these things and their applications will finally be shared in the end. For example, who – of the handyman, the surgeon, the researcher, and the manufacturer – will be entitled to what once the genetic test for melanoma has been marketed?

It is true that Anglo-American law, with its singular legal object of the trust, which allows the administration of the property of others in a spirit of confidence and based on the separation of property rights, could provide a pertinent model for considering the need for access. In French law, however, the trust exists only in a very different form, the trust agreement, designed for business and commerce (art. 2011 et seq. Civil Code). Moreover, we must assume that such a trust, at least in this context, is probably not that easy to implement, for it has not been mentioned in any of the English or American cases claiming property rights to human body parts and products.

A recent judgment from the US district court for the Southern District of New York, Association for Molecular Pathology v. U.S. Patent and Trademark Office (USPTO), 09cv4515,[15] illuminates the limits of property rights, including in its Common Law versions. At the same time it demonstrates the need to examine more deeply the question of access.

In this decision issued 30 March 2010, currently on appeal, the federal district court for Manhattan reached a radical decision concerning the patentability of body elements: that the patenting of human genes is contrary to American law, the "isolated DNA" patented not being markedly different from native DNA as it exists in nature. The two genes at issue are BRCA 1 and BRCA 2, mutations of which are responsible for some forms of ovarian and breast cancers. Myriad Genetics, an American firm, filed for a patent on the BRCA1 gene in 1994. Thereafter it either bought out or fought off any other company, laboratory, or hospital providing tests or performing research on the BRCA1 gene. By 1999 Myriad was the only company offering a test for and performing research on the BRCA1 gene. Myriad

[14]See Rifkin (2000).

[15]Available at: http://www.patentlyo.com/patent/2010/03/court-essentially-all-gene-patents-are-invalid.html (accessed 10 March 2011). See also *American Civil Liberties Union* website: http://www.aclu.org (then "Patent").

filed for a patent on the BRCA2 gene in 1995. The USPTO then granted Myriad exclusive rights over this gene. Following a similar and now standard policy on the patentability of human genes, the European Patent Office also granted protection to both genes and to the tests developed to identify mutations.

In the American case that concerns us here, the plaintiffs challenging the patents and their claims included women with cancer, women to whom screening had been recommended, public health and women's rights advocates, different colleges and associations of molecular biologists, pathologists, and geneticists together with individual physicians, geneticists, and counselors, and the Public Patent Foundation. The court held that the simple fact of purifying a natural gene cannot transform it in a patentable object. The transformed object must possess markedly different characteristics to satisfy the requirements of patentability.[16]

The key issue at stake was not really, however, whether natural phenomena could be patented: in this case, it is not a natural phenomenon that is patented, but indeed an invention, albeit routine, to identify a mutation and explain its role in the disease – none of which is "obvious". Moreover, the genes, removed from the donor, had to be worked on to become patentable and thus became commodities, thereby challenging at least in part the relationship with the person and weakening the argument often derived from genetic exceptionalism. According to this argument, widely challenged today, genes, because they express the biological uniqueness of the person, must be treated differently from other human body parts and products. It is on the basis of genetic exceptionalism, for example, that we protect genetic information most especially, because it is more likely than any other to provide a basis for discrimination. Patent law now denies some of this vision of things.

Indeed, the real question, therefore, was not the legal possibility of patenting human genes, but more specifically, the ability of property – in this case intellectual property – to reserve all use of these two genes to Myriad Genetics, to block other researchers seeking to develop a more reliable diagnostic test: Myriad's identifies only 80% of mutations, is very expensive and thus induces inequality among women (especially in a health care system, such as that in the US, where access to care is more difficult for the disadvantaged than in Europe). Those arguments, among others, were at the heart of these plaintiffs' claims: these patents create an obstacle to innovation, and finally, to the development and improvement of the patented diagnostic test. The relevant question was thus accurately formulated in terms of access – access to a particular material because it is necessary to the development of the research, and access to the uses derived from this material. We see here how the use of human body parts and products no longer leads to legal and theoretical questions of property, but inevitably instead to the political and practical questions of sharing.

---

[16]See 35 U.S.C. 101 Inventions patentable – Patent Laws. "Whoever invents or discovers any new and useful process, machine, manufacture, or composition of matter, or any new and useful improvement thereof, may obtain a patent therefore, subject to the conditions and requirements of this title".

From this perspective, this decision, although it may be subject to various criticisms, has the merit of underlining the inadequacy of the property paradigm for the exploitation of products from the body.

In sum, it is essential (more precisely, it was once essential) to demonstrate that the human body, its parts and its products, all of which manifest the sovereignty of their subject, can and must be thought of in law as things that are given, circulate, and are the object of diverse rights, patrimonial or otherwise. It is now essential to find legal rules that can take account of the fact that the human body, more than a thing, is a tangible and intangible complex, dissociable from the subject that shelters (or sheltered) it, in the service of diverse interests, not necessarily divergent, but that must be prioritized appropriately. Fundamentally pertinent when the issue is the initial donation by the subject (blood or tissue, for example), exclusive property rights constitute a cumbersome and imprecise basis for conceptualizing the further circulation of biological resources.

For a long time the person has obscured the body in the eyes of legal scholars; let us hope that an unfortunate intersection does not lead to property rights obscuring the questions of access and social justice.

References

Baud, J.-P. 1993. *L'affaire de la main volée. Une histoire juridique du corps*. Paris: Seuil.

Court of Appeals, R v Human Fertilisation and Embryology Authority, *ex parte Blood* [1997] 2 All ER 687, commented by Bergé, J.-S., "Droit communautaire, biomédecine et biotechnologies: entre concordance et antinomie", *RTDEur.*, 38 (4), oct-déc.2002.

Danos, F. 2007. *Propriété, possession et opposabilité*, p. 225, n° 204.préf. Laurent Aynès, Economica.

Hunter-Henin, M. 2001. *Pour une redéfinition du statut personnel*, Dissertation. Paris I, spéc. n° 146, p. 112.

Rifkin, J. 2000. *The Age of Access: The Revolution of the New Economy*. New York: Tarcher/Putnam.

Sefton-Green, R. 2000. «La procréation médicalement assistée entre droit national et droit communautaire. La controverse devant les cours anglaises ». *Revue générale de droit médical* (special issue): p. 101 et s.

Yearworth v N Bristol NHS CA. 2009. [2009] EWCA Civ 37.

Zenati-Castaing, F., and T. Revet. 2006. *Manuel de droit des personnes*. coll. Droit fondamental: PUF.

Chapter 14
Anonymity and Privacy in Biobanking

Judit Sándor and Petra Bárd

14.1 The Concept of Anonymity

Anonymity[1] is often seen as the best way to protect individual privacy in the biomedical context. The usual reasoning provides the following logic: data protection norms apply in case of identifiable data, however if the identity cannot be revealed no harm is done to anyone. With this connection of contentions anonymity becomes a defensive rather than protective element of personal privacy. However, the rationale behind data protection requires much more than simple escaping of legal problems. Furthermore, anonymity may often do harm, when an individual may no longer benefit of the findings of the research.

As early as 2001 Ellis and Mannion stated that the key to the permissibility of the use of genetic samples for research without consent is the anonymisation of genetic samples.[2]

But in many cases, by ensuring anonymity, privacy is excluded, as privacy is no longer there once no one can identify the origin of a biological sample or data. This is so despite the fact that before the concerns for genetic information in medical law (such as in the field of transplantation, blood donation), anonymity served entirely different functions. For instance, numerous cases in biomedical law prove that anonymity can serve as a guarantee for altruism, and in addition it is also seen as a safeguard against scientific and other biases in research. Donating blood or tissue to an identified person and to receive blood, tissue from an identified person creates a special and difficult interpersonal relationship between the donor and the recipient. The preferred method is an ultimate form of altruism, i.e. the total absence of a personal relationship, as in organ donation between complete strangers, where the identities of the donor and recipient are hidden through anonymisation.

---

[1] Anonymity is derived from the Greek word ἀνωνυμία, *anonymia*, meaning "without a name" or "namelessness". In colloquial use, anonymous typically refers to a person, and often means that the personal identity, or personal identifiable information of the given person is not known.

[2] Ellis and Mannion, (2001, 1).

J. Sándor (✉)
Central European University, 1051 Budapest, Hungary
e-mail: sandorj@ceu.hu

C. Lenk et al. (eds.), *Biobanks and Tissue Research*, The International Library of Ethics, Law and Technology 8, DOI 10.1007/978-94-007-1673-5_14,

In another context anonymity protects the individual against disadvantageous social consequences, such as exclusion from insurance. Therefore, while epidemiological data are necessary to plan health insurance schemes, the use of identifiable data is limited, for instance, for the purposes of quality assurance within health care or is justified in the field of private and commercial insurance.

In case of genetic and biobank research, anonymity has only a limited use as it is necessary to accompany DNA analysis with health care data to provide a meaningful conclusion. Therefore in most cases coded personal and health care data should be stored, not only physical samples. Furthermore, anonymity does not serve the interest of the donors as participants increasingly demand feedback on the findings relevant to their health care in biobank research.[3]

Anonymity as such is not the main focus of European data protection norms as it is interpreted and questioned only to see the limitation of data protection provisions but it is not defined positively. Still in the medical, scientific, and ethics literature anonymity has a principal role. Graeme Laurie distinguishes between "absolute" and "proportional" anonymity. While absolute anonymity is "achieved when no means are available to link data to an identifiable individual", we can talk of proportional anonymity whenever "no reasonable means of identification of specific individuals is possible,"[4] meaning linked or coded information, when the access to the link or the code is appropriately defined and restricted. Absolute anonymity should not be overestimated, since it might deprive researchers and donors from important values, e.g. it prevents longitudinal research being conducted or the feedback of results to research participants, still, the growing tendency to data mining at the same time shows the importance of at least some form of anonymisation. Many and often contradictory functions are associated with anonymity that should be re-examined in the case of biobanks.

In the following we will analyse anonymity in connection to privacy, confidentiality, health care and genetic research before we come to the functions of both anonymity and privacy in the legal framework of biobanks. After having highlighted the heterogeneity of norms and having offered certain technical solutions for anonymisation in a biobank context, we will explore two viable models: double coded samples in Estonia and the three-tier Hungarian solution. In the last part we will summarize our conclusions.

---

[3]See the findings of the second international workshop within the Tiss.EU project organized by Judit Sándor and Petra Bárd from the Center for Ethics and Law in Biomedicine (CELAB) at the Central European University (CEU), Budapest, Hungary. About 30 persons, speakers included, participated at the workshop that took place at CEU on 6–8 April 2009. The workshop made a major contribution to one of the four Focal Themes of the Tiss.EU project by addressing questions of "Anonymisation and Pseudonymisation as Means of Privacy Protection" (Focal Theme C) in relatively unexplored jurisdictions of Central and Eastern Europe, such as the Czech Republic, Hungary, Slovakia, and Romania. Due to the interdisciplinary nature of the workshop's subject, invited speakers represented a wide range of disciplines, such as law, medicine, philosophy, and information technology.

[4]Laurie (2002, 295).

14.2 Anonymity and Privacy

In comparison to anonymity, *privacy* is a much more dynamic notion and it serves slightly different functions. The core element of privacy is to maintain control over personal information which would be impossible once the data have been anonymized. Genetic data is never collected alone; in some jurisdictions dozens of pages long questionnaires need to be filled out by the patients or donors who often have to disclose special or sensitive information. On the one hand, these data are a treasure for researchers, on the other hand, they pave the way towards genetic or other types of discrimination. Should we attempt to overcome the drawbacks of deleting the link between the individual and his or her data, alternative means of privacy protection have to be found.

Personality can be protected by law in many different ways. Protection of dignity, liberty, autonomy, self-determination, privacy, or the right not to be discriminated against – these all serve diverse elements of personhood. While some rights such as the right to have someone's paternity acknowledged, or authorship, do require the use of the name and identity, there are some instances where personality is better protected by anonymity. For instance, *freedom of anonymous speech* is an important value in democracy: a person should be able to disseminate an opinion freely without disclosing his or her identity and without fear of retribution. The use of *pseudonyms* also protects personality. Women writers in Victorian times found it necessary to use male pen names to be taken seriously. For literature lovers, Mary Ann Evans is known to the world as George Eliot. Only a few people would know who Amandine Lucile Aurore Dupin was – but her pseudonym, George Sand, the first French female novelist of great reputation, is recognized by everyone.

In some cases, however, anonymity or anonymisation can be a harsh violation of rights – for instance, in the case of unrecognized authorship, when references are not used, or whenever paternity and identity are denied.

Issues of anonymity on the one hand and identity disclosure on the other hand are highly relevant in the contemporary debates of medical law as well. The anonymity of research subjects; the identity of gametes, tissues, and organ donors in cases of transplantation; the identity of gene donors in a biobank pose relatively novel concerns for bioethicists and scholars of biomedical law.

14.3 Anonymity and Confidentiality

Confidentiality "is the respectful handling of information disclosed within relationships of trust, such as healthcare relationships, especially as regards further disclosure. Confidentiality serves privacy. Researchers invariably promise to respect data-subjects' privacy, either by de-identifying the data to make them impersonal or by handling them securely."[5] While confidentiality originates from the

[5]Lowrance (2002, 8).

deontological norms of medical ethics, anonymity refers to a technical handling of data. The two notions are therefore strongly related to one another. While privacy has also an active dimension (control of personal information), confidentiality protects the doctor – patient relationship. The primary risks of a classical biobank to donors are related to the loss of confidentiality between either the doctor and the patient or the researcher and the donor. With time passing the nature of the danger normally diminishes to some extent, since research participants pass away and materials become archived ones. In biomedical research, in a biobank operated for therapeutic purposes or a population biobank, however the case is somewhat different: disclosing data against the data subjects' will, or accidentally identifying former donors (if the target group is small and research participants' identities are disclosed incidentally on the basis of circumstances even if information has been stripped of personal data that are mostly believed to contribute to identification) this may intrude well into the rights of persons other than the research subjects, first and foremost their relatives. This potential risk may be reduced by criminal[6] law sanctions, civil law sanctions,[7] or different forms of anonymisation. In order to avoid a breach of confidentiality, the same data protection rules and confidentiality standards have to apply for re-users of data, i.e. researchers in the original and in third countries. Due to the differences in the legal systems and the consequences attached to a breach of confidentiality, the safest way to transfer data to third countries is in anonymized format. As the Hungarian law of 2008 on the protection of human genetic data and the regulation of human genetic studies, research and biobanks (discussed below) prescribes, for the purposes of human genetic research, only anonymized, encoded or pseudonymised genetic samples or data may be transmitted to third countries, and only if the law of the given country provides for data protection corresponding to that under the Act No. LXIII of 1992 on the protection of personal data and the publicity of data of public interest. During the transmission of encoded genetic samples and data into third countries, the code key necessary for personal identification may not be transmitted.[8] An element of trust can be traced in the European – including the mentioned Hungarian – model,[9] since despite the discrepancies of the legal families of the European Economic Area, transfer of data to EEA countries shall be deemed as transfer within Hungary and the same confidentiality requirements are being presumed.[10]

[6] See for example Article 321 of the Swiss Criminal Code.

[7] See for example Article 33 of the Lithuanian Act on Data Protection on pecuniary and non-pecuniary damages.

[8] Article 28 (2).

[9] Article 29 Data Protection Working Party (2004, 19).

[10] Article 28 (1).

14.4 Functions of Anonymity in Health Care Law

As we saw one cannot tell whether anonymity is a positive value in itself in biomedical law: in some cases it has an important function, while in other fields anonymity can be a violation of important rights and interests. Ghost surgery (when surgery is conducted by someone who was not known by the patient prior to surgery) is a violation of informed consent. Publication of research results by disclosing relevant identifiers is a violation of privacy rights.

In the field of organ and tissue transplantation the name of the donor should be known to the medical staff. Moreover, a clear and accurate medical examination of compatibility is also an inevitable condition for donation while the recipient as a rule should not be connected to the (deceased) donors' family.

Clinical establishments involved in organ transplantations and coordinating organizations of transplantations do not disclose data and information in relation to donors and recipients, i.e. to those who are provided with organs. Under the current state of law organ transplantation is completely anonymous, and therefore all inquiries in relation to names or relatives are turned down. However, as in Hungary just a few such interventions are performed, if the donor's family declares that the organs of their relative will be used, the recipient may identify the donor from the scheduled date of the surgery. Furthermore, there exists no legal obstacle for the donor or the recipient to reveal this information.

The other field in medical law that kept anonymity as a main rule is the field of assisted reproduction when gametes are originated not from the social parents but from the donors. For a long time, in order to protect the integrity of the legal family, anonymity seemed to be a rule without exception. But if someone looks at the most recent changes in the field of the offspring's right to identity, it often seems to prevail over the donors' interests of anonymity. Of course, in the first countries where the laws on anonymity were changed, such as Sweden and United Kingdom, the legislative power did not adopt laws with retroactive effect. After the entry into force of the relevant legal instruments, the donors must be informed in advance on the possibility to reveal their identity in front of the child when it reaches maturity.

Having highlighted the meaning and importance of anonymity in related fields, in the following we will focus on anonymity in genetic research and anonymisation of data in genetic biobanks.

14.5 Genetic Research

Issues of anonymisation came into the frontline of the literature with the spread of large scale genetic tests and human genetic research. Human genetic research, being engaged with the structure of genetic material (genes and chromosomes), their disorders and the appearance in physical, intellectual, and behavioural features of the genetically encoded programme and the regular features of the transmission from the parents to the offspring of the genetically encoded programme and the exploration of the disorders of these processes, has an overwhelming scientific significance. The laws on genetic research lay down the framework of the use for research of samples and data, data protection guarantees necessary for use, rules on genetic research on the population and the conditions of samples in the archived collection for a new research.

If the genetic sample taken in the context of genetic testing is intended to be used for research purposes, a repeated consent procedure is required by the law.

Genetic research on human behaviour should be conducted in a fashion that respects the dignity of the research participant by taking into account not only genetic but also the extra-genetic features of the personality. At the moment there are no detailed provisions how to ensure this. One solution could be if a social scientist having knowledge in psychology or/and sociology were involved in the study in order to avoid stigmatisation of the research participant and with the aim to help in developing a more balanced assessment of personality. By this method the danger of genetic determinism and reductionism could be more easily avoided.

During the research, the person concerned may request the encoding, pseudonymising or anonymising of the genetic sample intended for research purposes and that of the derived genetic data. The fate of the data in a biobank, its form of anonymisation, or possible destruction thus depends on the data subject – at least until a link can be established between him or her and the information or the sample. However simple that may sound, the complexity of the issue is highlighted by the fact that we do not possess a common definition of crucial terms, such as 'biobank' and 'anonymisation'. In the following we will give a brief overview of these notions.

14.6 Anonymity in Biobanks

DNA sampling, data collection, sharing and exchange of information are all important for genetic research, clinical care, and future treatments. However, the corresponding ethical and legal framework is not well defined. Most health care institutions have no written policies or agreements regarding this activity, and even if there was a willingness on the side of hospitals, clinics, and research institutes to adjust their practice to some general norms, researchers or drafters of these internal guidelines are in an extremely difficult position due to the large number of international, national, and professional guidelines that contain different, sometimes even contradicting recommendations relevant for biobanks.

A fundamental underlying question is how we define biobanks. Repositories of human samples and related data can be grouped along the stored material, which can be organs, tissue, blood, cells or other materials, such as urine or liquor. Biobanks can also be distinguished according to their sizes: these repositories may vary from population biobanks to three samples in a pathologist's refrigerator. A biobank does not only contain human biological samples, data are also stored there. Robert F.

Weir and Robert S. Olick define biomedical classical (clinical) research databases as follows: a database that is developed and restricted to authorized clinical investigations (e.g. oncology, pathology, etc.) in several academic medical centres.[11] These databases contain genetic and other biomedical information about individual patients, derived from their clinically collected tissues, with the electronic data sometimes being transmitted to a central database. The above mentioned two scholars differentiate between commercial databases, which are human tissue databases that are restricted to scientists willing to pay to have access to DNA sequences and the databases that include other protected information.

A population biobank, based on the definition of the Council of Europe[12], is a collection of biological materials that has the following characteristics: (i) the collection has a population basis; (ii) it is established, or has been converted, to supply biological materials or data derived there from for multiple future research projects; (iii) it contains biological materials and associated personal data, which may include or be linked to genealogical, medical and lifestyle data, and which may be regularly updated; and (iv) it receives and supplies materials in an organized manner.

Forensic databases greatly differ in nature from the above classical and population biobanks. In the broad sense forensic databases are DNA databanks held by authorized laboratories of police and official forensic institutions for criminal and other legal procedures, such as the identification of victims, missing persons, perpetrators, the establishment or rejection of paternity, etc.

One may think of other divisions of biobanks as well, but the crucial point for our current discussion is the double nature of these databases, i.e. the fact that they contain both tissues and data, that is information on the donated human biological material and the donor linked to these tissues. Therefore the question arises of whether traditional data protection rules are an effective tool in the fight against the misuse of information, and whether anonymisation of samples is the best safeguard, or on the contrary, whether it limits the autonomy of research participants in a biobank – a question very much related to the issue of genetic exceptionalism.[13] Some authors[14] state that since tissues and data are different, they raise different issues. Therefore different regulations are said to be needed. At the same time, there

[11]Weir and Olick (2004, 294).

[12]Recommendation Rec(2006)4 of the Committee of Ministers to member states on research on biological materials of human origin, Article 17.

[13]The *Genomics Law Report* defines genetic exceptionalism in the following way: "Genetic exceptionalism is the concept that genetic information is inherently unique and should be treated differently in law than other forms of personal or medical information. There are several reasons for such special consideration: genetic information can predict disease occurrence in a person and their blood relatives; it uniquely identifies a person; and it can be used to discriminate and stigmatize individuals. While these issues deserve attention and steps should be taken to protect people, over-regulation could limit our ability to investigate how genetic information predicts disease and improve medical outcomes." Available at www.genomicslawreport.com/index.php/2009/10/06/genetic-exceptionalism-and-the-precautionary-principle (accessed March 11, 2011).

[14]Trouet and Sprumont (2002, 3–19).

is a trend for unified regulations as well. If DNA represents special human rights questions, its protection should reflect these corresponding concerns.

Many scholars and researchers consider tissue research a form of medical research, or at least realize the similarities between the two, and therefore propose that the protection of personal medical data shall cover this field. Confidentiality can be ensured through various legal means: antidiscrimination laws prohibit discrimination, while criminal laws may also sanction discriminatory behaviour. Confidentiality may also be ensured through anonymisation. This is the point where scientists' interests may clash with legal requirements. Based on the study "Ethical and regulatory aspects of biobanks: global consensus and controversies", Bernice Elger summarized the literature and regulatory frameworks on confidentiality, anonymisation and consent.[15] Elger and her colleagues interviewed persons from all related disciplines, such as scientists, biobankers, physicians, lawyers, and ethicists from different parts of the world and from different types of institutions. Experts and researchers agreed on only a few issues: first, they are opposing irreversible anonymisation of samples at the time of collection and storage; second, in their view, researchers have to be tightly controlled; third, a distinction needs to be drawn between publicly and privately funded projects; fourth, it is strongly advisable to place research data and results in the public domain within a reasonable time-frame in order to stimulate scientific progress; and finally, they call for a unified definition of anonymisation and establishing common conditions under which material and data are shared with others. The last issue is especially topical for our analysis, since data sharing and transnational research are hampered by the differing understandings of anonymisation and pseudonymization.

Pseudonymisation refers to a technique of data processing in which anonymity is assured assure while keeping a link to be able to update information and to re-contact participants whenever information valuable to the donors is discovered. The next logical step is to determine what kinds of pseudonymisation techniques are adequate: double coding, single coding or some other method. Even if one term refers to a certain technique method, a lack of consensus on the normative definition prevents researchers from agreeing on standardisation.

14.6.1 Heterogeneity of Norms

In the myriad of terms one can find references to samples that are anonymous, anonymised, anonymously coded, coded, unidentified, de-linked, permanently de-linked, not traceable, unlinked, identifiably linked, pseudonymised, encoded, encrypted, directly identified, confidential, identifiable, not traceable, or in the

[15] Ibid. Other collaborators in the research were: Nikola Biller-Andorno, University of Zurich, Switzerland; Agomoni Ganguli-Mitra, University of Zurich, Switzerland; Andrea Boggio, Bryant University, USA; Alexander Mauron, University of Geneva, Switzerland; and Alexander M. Capron, University of Southern California, USA.

UNESCO terminology[16]: data linked and unlinked to an identifiable person, furthermore, data irretrievably unlinked to an identifiable person.[17] Data unlinked to an identifiable person means data replaced by or separated from all identifying information about that person by use of a code. It can be applied to a biological, whereas data irretrievably unlinked to an identifiable person is data that cannot be linked to an identifiable person, because the link to any identifying information has been destroyed.[18]

Different legal families adhere to distinct legal traditions, and prefer one or another term over others for legal historical reasons. Sometimes even the same term is used with a different meaning, like the words "anonymised" and "coded" which are filled with different content in continental and common law jurisdictions.

In the European setting the right to privacy is laid down in Article 8 of the European Convention for the Protection of Human Rights and Fundamental Freedoms. The European Court of Human Rights deducted the right to data protection from that provision. It regularly refers to the 1981 Convention for the Protection of Individuals with regard to Automatic Processing of Personal Data (ETS No. 108) and the Additional Protocol to Convention 108 regarding supervisory authorities and trans-border data flows (ETS No. 181), also adopted in the framework of the Council of Europe. Among the European standards, Recommendation Rec (2006) 4 of the Committee of Ministers of the Council of Europe to member states on research on biological materials of human origin can be referred to first. The instrument[19] distinguishes between non-identifiable and identifiable samples. The former are unlinked samples, while the latter are linked anonymised and coded samples.

As to European standards on anonymisation specifically, Recommendation No. R (97) 5 of the Committee of Ministers to Member States on the Protection of Medical Data may be helpful: "the expression 'personal data' covers any information relating to an identified or identifiable individual. An individual shall not be regarded as 'identifiable' if identification requires an unreasonable amount of time and manpower. In cases where the individual is not identifiable, the data are referred to as anonymous" (Principle 1).

Apart from the Charter of Fundamental Rights, and more specifically its Articles 7 and 8 on privacy and data protection, the main European Union data protection instrument is Directive 95/46/EC. This document, however, does not speak of an "unreasonable amount of time and manpower," but states in Article 2a that

> "personal data" shall mean any information relating to an identified or identifiable natural person ("data subject"); an identifiable person is one who can be identified, directly or indirectly, in particular by reference to an identification number or to one or more factors specific to his physical, physiological, mental, economic, cultural or social identity.

---

[16]International Declaration on Human Genetic Data, 16 October 2003, Article 2, Points (ix) and (x).

[17]Elger and Caplan (2006, 661–66).

[18]UNESCO International Declaration on Human Genetic Data, 2003, Article 2, (x) and (xi).

[19]See Article 3 on the identifiability of biological materials.

In the context of biobanks, especially when it comes to information security, Article 17 of the Directive is worth deeper exploration. According to this provision the controller is obliged to

> implement appropriate technical and organizational measures to protect personal data against accidental or unlawful destruction or accidental loss, alteration, unauthorized disclosure or access, in particular where the processing involves the transmission of data over a network, and against all other unlawful forms of processing.

Article 29 of the EU Data Protection Directive 95/46/EC establishes a "Working Party on the Protection of Individuals with regard to the processing of Personal Data" (hereinafter referred to as the "Article 29 Working Party"). According to the Article 29 Working Party, electronic health records create a new risk scenario,[20] and acknowledging that genetic data may pose special risks even among sensitive data, the data controller can be requested to carry out risk assessment, establish security policies and provide ongoing training for staff.[21]

Bernice Elger and Arthur Caplan summarize the European approach to distinguishing levels of anonymisation in the following.[22] The first category of *anonymous* DNA samples does not exist, only for "archaeological" tissue for which no material for comparison to an identified person exists. The second type of samples are *anonymised* ones, which are stored alongside certain information which is crucial for research, but information that would allow the identification of the donor is all stripped. Depending on whether the latter information can be restored or not, anonymised samples can be divided into *irreversibly anonymised* (unlinked) and *reversibly anonymised* (linked) ones. In the latter case identification is possible via a code (pseudonym), but researchers or users of the material do not have access to the code. *Coded* samples are like linked (reversibly) anonymised ones, with the difference that researchers or users do have access to the code. One has to be cautious with the terminology, as in the US "anonymised" means irreversibly unlinked or reversibly linked, but the researchers do not have access to the code, while in Europe the word "coded" means reversibly linked, where researchers have or do not have access to the code. The last category is that of *identified samples*, where the information stored along the samples permits the direct identification of the donor, such as when the name and birthday are indicated on a tube.

Putting these terminological discrepancies apart, the main controversy has evolved around the question of how to assure adequate anonymisation – be it linked or unlinked. Who shall decide which degree of anonymisation is adequate? How many characteristics must be stripped to obtain truly irreversible or reversible economisation? What are the standards for technical questions of security?

[20] Article 29 Data Protection Working Party (2007a, 11).

[21] Article 29 Data Protection Working Party (2004, 11–12).

[22] Elger and Caplan, *op. cit.*

14.6.2 Technical Solutions of Data Security

Addressing technical questions of security, the need for standards and for cooperation with IT institutions has been stressed. Again Article 29 Working Party may give some guidance as to the preferred standards: in its Opinion 4/2007 the Working Party stated that "Pseudonyms should be random and unpredictable. The number of pseudonyms possible should be so large that the same pseudonym is never randomly selected twice. If a high level of security is required, the set of potential pseudonyms must be at least equal to the range of values of secure cryptographic hash functions."[23] In its previously drafted working document on the processing of personal data relating to health in electronic health records[24] Article 29 Working Party promoted, among diverse technical solutions, Privacy Enhancing Technologies PETs. PETs are IT solutions that mitigate the drawbacks of technological development in personal data management, so that donors (or data subjects in general) regain influence over information about themselves. The Working Party also proposed that legal safeguards refer, among others, to the development of a reliable and effective system of electronic identification and authentication as well as constantly updated registers for authorized persons who can access databases; documentation of all processing steps which have taken place within the system; and preventing unauthorized access or alteration of data.[25]

When searching for effective anonymisation in case of biobanks, one may borrow solutions from other fields where data protection is a concern, such as protocols in securing internet communications, emails, online purchase, etc. A viable solution is the anonymous tracking model for individual minority subsidies, where the state wishes to reduce or eliminate the disadvantages suffered by certain minorities by positive discrimination or affirmative action programs. An interesting field for comparison is the case of minority subsidies. The question is how to subsidise disadvantaged minorities if we cannot identify them, because the law prohibits having certain characteristics (e.g. ethnicity) registered. On the one hand, these pieces of sensitive information enjoy special protection, and we wish to deny authorities' access to it. On the other hand, this information would be crucial in order to have an effective and correct system of subsidy, free from abuse and embezzlement. In order to overcome the problem, several authors propose the use of modern information technology, unidirectional data transformation procedures, and emphasize the importance of trusted third parties. The technique can be adapted to different settings, like genetic databanks, as well, especially since in both cases the handling of sensitive, special, classified information is at stake. In such a sensitive setting one may make use of the available modern information technology which offers data

---

[23] Article 29 Data Protection Working Party (2007b, 18).

[24] Article 29 Data Protection Working Party (2007a, 11).

[25] Ibid., 19–20.

management technology, that allows to make the link between the data and the data subject, in our case the patient, the donor, or the suspect, unidirectional.[26]

A probabilistic distortion method was suggested by Johannes Gehrke from Cornell University, which would totally disable re-identification of donors. According to this method an extremely small probabilistic number (that might be positive, negative, or zero) is added to the values in the database, thereby distorting the original figures so they can never be traced back. This number can be mathematically tailored, customized, so that the statistical properties of the attributes may be the same, i.e. the probabilistic number is small enough not to modify the outcome of the research, but is large enough to ensure that data cannot be joined by simply testing equality attributes. Of course, this system, just as any other alternatives, may have some drawbacks: it may lead to distorted results in case specific attributes are being compared, and it clearly introduces an uncertainty element.[27]

The fact that a sample is seen as being anonymised has vital consequences from the point of view of obligations of acquiring informed consent. Should the definition of anonymisation be broadened to too many types of pseudonymised samples, or, if the definition is softened by reference to a minimal risk of identification, or to reasonable effort, or reasonable amount of time and manpower needed for identification, data protection rules do not apply any more. More specifically, in the EU context we can derive from Recital 26 of Directive 95/46/EC that data collected in an anonymous way, or data that have been rendered anonymous at a later point in time, are outside the scope of the Data Protection Directive 95/46/EC, since "the principles of protection must apply to any information concerning an identified or identifiable person" only. Therefore, the British approach prescribes safeguards even for anonymisation, since thereafter, on the one hand, the data subject will not be able to influence the fate of the data relating to him or her, and on the other hand, legal guarantees will not apply. Therefore, the Office of the Information Commissioner issued a legal guidance on the Data Protection Act of 1998, which states that "in anonymising personal data the data controller will be processing such data and, in respect of such processing, will still need to comply with the provisions of the Act."[28] Thereafter however – in line with the provisions of the Directive – the data fall outside the scope of the law.

14.6.3 The Case of Estonia: Double Coded Samples

Current Estonian informed consent rules have their roots in the European Union Data Protection Directive 95/46/EC. However, the European Union rules do not

[26] See, for example, Claerhout and DeMoor (2005, 257–65); Székely (2009, 27–62).

[27] Referred to by Zoltán Alexin at the second international workshop organised in Budapest within the Tiss.EU project. See also Xiao, Wang, and Gehrke (2009).

[28] Data Protection Act 1998 (2002, 15).

include extensive informing requirements; consent sometimes is not even necessary. Informed consent in data protection differs greatly from informed consent in medicine. Informed consent in medicine (i) aims at the protection of life and health, (ii) extensive information is required, (iii) it is project specific, and (iv) consent is almost always necessary. In comparison, informed consent in data protection (i) aims to protect privacy, (ii) less information is required, (iii) specification of one field of use is enough, and (iv) there are a number of cases where consent is not necessary.[29]

In the Estonian Genome Project data is stored in a coded form, where persons are identifiable. In case of a consent withdrawal, the code will be erased; however, erasure of all data can also be applied for. Whenever needed, data is issued in pseudonymised form from which data subjects cannot be identified – neither directly, nor indirectly. This is realised through the so-called five donors rule, which ensures that every data in the database matches at least five persons. For each and every research use the ethics committee's approval is needed.

The previous consent form is not suitable for the Estonian Genome Project, since it enables future research with yet unknown project goals. Theoretically one could opt for asking for a specific consent at a later stage after data collection. However, according to the Estonian expert, for practical reasons, taking into account the nature and level of risks, considering autonomy as empowerment rather than as a disempowerment, bearing in mind the value of biological samples, population biobanks deserve a new type of informed consent. Therefore, in the Estonian Genome Project an open consent requirement has been adopted. Open consent is consent to participate in a population biobank and in research projects utilising data and/or samples from a biobank. The consent is open, i.e. not limited in respect of time, projects, researchers, etc. It justifies interference with bodily integrity and data privacy. It should be noted, however, that even open consent does not give authorization for everything. Neither is it an indication for commitment to future participation, i.e. withdrawal of consent is still possible. Further, the open consent system relies on certain conditions, like public control.

Concerning the crucial issue of withdrawal of the gene donor's consent, Article 12 Section (4) point (7) of the Human Genes Research Act (HGRA)[30] sets forth that gene donors have the right to withdraw their consent until tissue samples or the descriptions of state of health are coded, and in such case the gathered information and blood sample shall be destroyed. Afterwards, a gene donor has the right to apply, at any time, to the chief processor for the destruction of data that enables decoding.[31] However, destruction of data that enable decoding (i.e. anonymization) does not mean also the destruction of biological material or other data.[32] In line

[29]See the findings of Ants Nõmper at the second international workshop organised in Budapest within the Tiss.EU project.

[30]Human Genes Research Act (passed by the Riigikogu) (December 13, 2000, 104, 685).

[31]Article 20 Section (1) HGRA.

[32]Nõmper and Kruuv (2003, 213–24).

with Article 10 Section (2) of the HGRA, a gene donor has the right to request termination of biological material and other data available in the genebank, if the identity of a gene donor is unlawfully disclosed. After the data that allows decoding is destroyed, the health data and the tissue samples of a gene donor, stored in the genome bank, are anonymous. Thus, the regulation of personal data protection does not apply, since it does not extend to data processing performed with anonymised data in line with Article 7 Section (2).

14.6.4 The Case of Hungary: Three Options to Grant Confidentiality of Genetic Samples

The Hungarian law adopted in 2008 on the protection of human genetic data and the regulation of human genetic studies, research and biobanks presents a unique solution on the European continent, therefore, we shall elaborate the details of the regulation. First, the debates surrounding lawmaking will be presented and second, we will discuss the rules on anonymisation in greater detail.

Hungary became Party to the Oviedo Convention for the Protection of Human Rights and Dignity of the Human Being with regard to the Application of Biology and Medicine of 4 April 1997, and to its Additional Protocol on the Prohibition of Cloning Human Beings by Act VI of 2002. Thereby Hungary undertook to monitor the regulatory range of bioethics and medical-biological research continuously and to prepare legislation in this subject, and that is what the Act aims to comply with.

As to European Union legislation, the following was taken into account by the lawmaker: Directive 95/46/EC of the European Parliament and of the Council of 24 October 1995 on the protection of individuals with regard to the processing of personal data and on the free movement of such data and Directive 2004/23/EC of the European Parliament and of the Council of 31 March 2004 on setting standards of quality and safety for the donation, procurement, testing, processing, preservation, storage, and distribution of human tissues and cells.

The Parliamentary Assembly of the Council of Europe discussed the draft Additional Protocol on genetic testing for health purposes to the Oviedo Convention, covering the field related to the Act, with the exception of genetic research, at its session in Strasbourg between 21 and 25 January 2008. During the preparation of the Hungarian Act on biobanks, the draft of this Protocol was also reviewed.

As a result of the legislative process, Act No XXI of 2008 on the protection of human genetic data and the regulation of human genetic studies, research and biobanks entered into force on 1 July 2008. The law has become shorter and simpler than it was foreseen in the original policy paper in a desperate effort to avoid sensitive issues. Thus, the law addresses the use of genetic information only in a very narrow biomedical sector: in the fields of genetic testing, screening, and research.[33]

---

[33] In Article 1 of the Act the purpose of the law is stated as "to lay down rules on human genetic tests and screening (studies) and human genetic research, the conditions and purposes of the treatment of

The law restricts the use of genetic data only in this biomedical context, so even in the lack of regulation of the broader use of genetic data based on the Act genetic data processed for diagnostic or research purposes cannot be disseminated for the purposes of insurance. Despite the intended laconic law, the mere word "genetics" was an invitation for a vehement debate by various political actors. Fears of genetic discrimination, exploitation or trafficking data to foreign countries were the major concerns in the political debate.

Even earlier in the course of the legislative debate lawyers, data protection activists were mobilized and advocated for newer and newer guarantees for the protection of genetic data.[34] The issues of data protection were so dominant in the debate that some other, broader human rights aspects were entirely left out from the final version of the law.[35]

By focusing on data protection questions, such as how to store genetic data (should it be stored as anonymous data, coded, single or double coded, and who should get access to the code or who should keep the code?) and creating a stronger protection for the genetic data, some other elements of the ethical-legal framework were sacrificed, such as the prohibition of discrimination based on genetic characteristics; they were referred to general laws. As relatively little public consultations were conducted on biobanks in Hungary, various data protection rules, including the protection of the health care data are fragmented and dispersed in various norms.

The consequences of the careful approach towards data protection make the Hungarian solution unique. The specificity of the Hungarian Parliamentary Act in its final form is that it regulates three different levels of coding and anonymity:

(a) *the encoded genetic sample or data* means genetic sample or data regarding which all the personal identification data relating to the person giving the sample are replaced by a code;
(b) *pseudonym genetic sample or data* means encoded genetic sample or data regarding which the code replacing the personal identification data was provided to the person concerned;
(c) *anonymised genetic sample or data* means genetic sample or data regarding which all the personal identification data relating to the person giving the sample was made incapable of identifying the person.[36]

---

genetic data and rules on biobanks." The Act applies to genetic sampling for human genetic study and human genetic research performed under this Act in the territory of the Republic of Hungary, the processing of genetic data irrespective of the place of sampling, and to genetic testing and screening and human genetic research and to biobanks.

[34]In order to understand the main focus of the debate, it should be mentioned that in the Hungarian law, the most considerable field within the right to privacy is the protection of personal data. The classical concepts of inviolability of domicile and secrecy of correspondence are also important subjects to be protected, but there is a much higher uncertainty in the abstract fields of privacy e.g. concerning the right to disposal of someone's personal body.

[35]Such as the right not to be discriminated against in the field of public health insurance and education.

[36]See Article 3 Points (d), (e), and (f).

For a long time it was believed, also in Hungary, that anonymous data could guarantee the highest level of protection for the genetic data. However, many problems were identified in respect of systematic anonymization of genetic samples and data. First of all, anonymous data cannot be matched with other health data, and as such the relevance of the data is reduced for scientific research. Anonymisation is also contestable, taking into account the participants' interests, since a further feedback, based on the request of the owner of the sample, would be usually impossible.

The third part of the law needs greater elaboration, as it provides specifically for rules on the operation of *biobanks.* By legislating on the operational rules of biobanks, the conditions of the operation of collections containing human biological material samples shall be established. Accordingly, the genetic samples and data shall be stored only in biobanks and, as a general rule, in a format determined by the declaration of consent of the person concerned. There is a safeguard provision providing for the conditions of the storage of the genetic sample or data in a way that allows personal identification and states the prohibition of a register involving information that contains personal identification data. It is stated that a biobank may be established and maintained by a health service provider authorised to conduct genetic studies and certain medical researchers and another institution entitled to conduct human genetic research only.[37] Larger scale population based study is also mentioned in the Act. Under Article 17 human genetic research on the population may be conducted for the determination of the distribution of genetic variants between individuals within a given group or between individuals belonging to different groups, and to the exploration of the nature and consequences of the latter. The law provides for the tasks of the responsible person being employed in the biobank, the keeping of data stored in the biobank and the forwarding of data as well as the register of biobanks.

During the storage of the genetic sample or data, the protection of these shall be ensured against destruction, termination, change, injury, publication or access by unauthorised persons.[38] Unless provided otherwise by this Act, genetic samples and data shall be stored in an encoded format. Encoded genetic samples, data and code keys shall be stored separately, both physically and electronically. Access to the code key shall be authorised to a person being responsible[39] within the framework of the Act. During the separate storage of the code key, it shall be ensured that no other person may access it apart from the person entitled thereto. The code of the pseudonym sample or data shall be put at the exclusive disposal of the person providing the sample. Storage of genetic sample or data together with personal identification data is

[37] Article 23 (1).

[38] Article 24.

[39] Within the biobank, the person responsible for the protection of genetic samples and genetic data, the registering of genetic samples and data and the keeping of the register shall be the head of the institution maintaining the biobank and the person designated by the latter for the supervision of the operation of the biobank. Article 26 (1).

subject to the consent of the person concerned.[40] A register containing genetic samples and data stored together with personal identification data or encoded genetic samples and data may not be linked to a register containing personal identification data.[41]

Every genetic sample and data stored in the biobank and all related procedures and activities and the forwarding of the genetic sample and data shall be registered for at least 30 years following the recording of the data, except when the person concerned withdraws his or her consent to the treatment of genetic data. In such a case, every register relating to genetic data shall be destroyed following the information of the person concerned. The register shall contain the types, quantities, origins and destination of collected, studied, stored, processed and distributed or otherwise used genetic samples and the genetic data derived from these. After expiry of the mandatory registration period, the data shall be subject to treatment under the Act XLVII of 1997 on Health Care Data.

14.7 Conclusions: End of Anonymity?

In case of contemporary biobanks legislations it seems that we have much less emphasis now on anonymity than at the dawn of the first biobanks. Anonymity seems to be an illusion that we might never be able to achieve in case of genetic samples, and perhaps it no longer serves the interests of research participants. Anonymisation was seen as an attractive tool in the securing privacy and autonomy, in the prevention of harm that the leaking out of information to unauthorized persons – such as insurers or employers – may mean. Securing privacy, confidentiality, data protection or autonomy does not however mean full anonymization. First, because anonymous DNA samples do not exist, since theoretically one might always derive samples from living donors and archived materials, and compare them to a sample in a biobank. Second, and more importantly, anonymisation does not necessarily serve the interest of donors. Donors shall be aware of the details of data protection, such as: who has access to their data? Under what conditions do they have access, and what is the level of security?[42] Patients cannot determine the destination of their samples. Furthermore, feedback of research results is the most desired outcome of such research, which is also disabled by delinking samples and data. Thirdly, stripping the code or delinking information from samples prevents researchers from going back to patients and conduct longitudinal research. Ultimately, by slowing down research, anonymisation may harm donors and non-donor patients more than a secure system of pseudoanonymisation. Such a system, of course, cannot operate without an element of trust, which can only be established if the necessary institutions, safeguards (including confidentiality

---

[40]Article 25 (1).

[41]Article 25 (2).

[42]Chadwick (2001, 203–10, at 207).

requirements, respecting donors' and their relatives' privacy and the right not to know, etc.), and procedures are created – not without establishing a corresponding legislative framework.

References

Article 29 Data Protection Working Party. 2004. Working Document on Genetic Data of 17 March 2004, 12178/03/EN, WP 91.

Article 29 Data Protection Working Party. 2007a. Working Document on the Processing of Personal Data Relating to Health in Electronic Health Records (EHR) of 15 February 2007, 00323/07/EN WP 131.

Article 29 Data Protection Working Party. 2007b. Opinion 4/2007 on the Concept of Personal Data of 20 June 2007, 01248/07/EN WP 136.

Chadwick, R. 2001. "Informed Consent in Genetic Research." In *Informed Consent in Medical Research*, edited by L. Doyal and J. S. Tobias, 203–10. London: BMJ Books.

Claerhout, B., and G. J. E. DeMoor. March 2005. "Privacy Protection for Clinical and Genomic Data: The Use of Privacy-Enhancing Techniques in Medicine." *International Journal of Medical Informatics* 74 (2): 257–65.

Elger, B. S., and A. L. Caplan. 2006. "Consent and Anonymization in Research Involving Biobanks: Differing Terms and Norms Present Serious Barriers to an International Framework." *European Molecular Biology Organization Reports* 7 (7): 661–66.

Ellis, I., and G. Mannion. 2001. "Humanity Versus Utility in the Ethics of Research on Human Genetic Material." *Genetics Law Monitor* 1 (5): 1.

Human Genes Research Act (passed by the Riigikogu). December 13, 2000. (as RT I 2000, 104, 685), entered into force on January 8, 2001 (RT is *Riigi Teataja* or *State Gazette*). In the original language: *Inimgeeniuuringute seadus*, official English translation is available at http://www.legaltext.ee/text/en/X50010.htm. Accessed March 11, 2011.

Laurie, G. T. 2002. *Genetic Privacy: A Challenge to Medico-Legal Norms*. Cambridge: Cambridge University Press.

Data Protection Act of 1998. 2002. *Legal Guidance*. London: Office of the Information Commissioner.

Lowrance, W. W. 2002. *Learning from Experience: Privacy and the Secondary Use of Data in Health Research*. London: Nuffield Trust.

Nõmper, A. and K. Kruuv. (2003) "The Estonian Genome Project." In *Society and Genetic Information: Codes and Laws in the Genetic Era*, edited by J. Sándor, 213–24. Budapest: CEU Press.

Recommendation Rec (2006) 4 of the Committee of Ministers to member states on research on biological materials of human origin.

Székely, I. 2009. "Positive Discrimination and Data Protection: A Typology of Solutions and the Use of Modern Information Technologies." In *Privacy Protection and Minority Rights*, edited by M. D. Szabó, 27–62. Budapest: Eötvös Károly Policy Institute. Accessed March 11, 2011. http://www.ekint.org/ekint_files/File/kiadvanyok/privacy_minority.pdf.

Trouet, C., and D. Sprumont. 2002. "Biobanks: Investing in Regulation." In *Baltic Yearbook of International Law*, edited by I. Ziemele, 3–19, vol. 2. Leiden: Brill/Martinus Nijhoff Publishers.

UNESCO International Declaration on Human Genetic Data, 16 October 2003.

Weir, R. F., and R. S. Olick. 2004. *The Stored Tissue Issue*. Oxford: Oxford University Press.

Xiao, X., G. Wang, and J. Gehrke. 2009. Interactive Anonymization of Sensitive Data, *Proceedings of the 35th SIGMOD International Conference on Management of Data, 2009*, 1051–1054. Accessed March 11, 2011. www.cs.cornell.edu/bigreddata/publications/2009/sigmod2009-p1051-xiao.pdf.

Epilogue

As we have seen in cases of tissue collections we witness a heterogeneity of various ethical and legal terms and approaches. In many jurisdictions tissue collections and biobanks exist under different names. Various overlapping terms have been in use, including the terms "registries, repositories, biological archives, pathological sample-collections, genome databases, gene-banks, population biobanks" etc. While some years ago the term "biobank" was almost unheard of, it seems to be used too broadly now. Therefore, the crucial moment for adopting a European-wide definition for biobanks potentially has already passed. Not just the wide application of the catchy term "biobanks" but also the controversies in the social meaning of genetic data and some countries' rejection of the term "bank" (because of the strong connection with the commercial applications) seem to be obstacles to the late harmonisation of the term "biobank". Furthermore, it seems that while biobanks for genetic studies were the centre of attention about four or five years ago, now new forms of biological collections also have to be taken into account. Particularly in the field of regenerative medicine, tissue and cell collections have been developed for purposes of research and therapy that aren't genetic in nature.

Framing and reframing new technologies under the EU Tissue Directive[1] and the EU Regulation on Advance Therapies[2] seems to even further complicate the question. Clinicians often misunderstand the scope and applicability of these norms, in addition to the fact that there are countries that tend to apply the norms of organ and tissue transplantation even in the field of tissue research in biobanks. In general, it can be noted that while the subject matter – such as human tissue – involves various activities for clinicians and researchers, from harvesting and collecting, to therapeutic use and research, in law, the function of the tissue determines the legal framework. It follows that the use of tissue for human therapy or in vivo research requires strict safety measures, while doing in vitro research on human tissues raises

[1]Directive 200423/EC of the European Parliament and of the Council of 31 March 2004 on setting standards of quality and safety for donation procurement, testing, processing, preservation, storage and distribution of human tissues and cells.

[2]Regulation (EC) No 1394/2007 of the European Parliament and of the Council of 13 November 2007 on advanced therapy medicinal products and amending Directive 2001/83/EC and Regulation (EC) No 726/2004.

C. Lenk et al. (eds.), *Biobanks and Tissue Research*, The International Library of Ethics, Law and Technology 8, DOI 10.1007/978-94-007-1673-5,

other types of issues, mainly involving data protection and consent. In the field of tissue collections for in vitro research, and more specifically, for biobank research, the European approach almost exclusively singles out the data protection norms, even if many other aspects of tissue banking would deserve regulatory attention.

So how should this epistemological complexity be tackled? One possible way could be the systematic and regular mapping of the often-changing legal framework by the European Group on Ethics in Science and New Technologies (EGE) in consultation with National Ethics Committees. Such European co-operation could answer the question of whether new technologies in the pipeline fit into the already-existing framework, or require some adjustment. While we think it is inevitable that different disciplines, such as ethics, law and medicine, apply different terms, legislators and regulatory bodies should be aware of these differences in order to minimise the danger of the ambiguous interpretation of regulatory instruments, namely the reading of one instrument in two different ways. We noticed that between two extreme positions – one that ignores new technologies and the other that adopts legal standards in a hasty fashion – regulators should make all possible efforts to find ways to create a coherent ethical and legal framework.

After the field has been regulated by numerous fragmented ethical and legal instruments, a possibility of a broader instrument could be compiled that covers the entire field coherently, possibly every five to ten years. The description of this situation shows the importance of international and overarching initiatives like the P3G project (www.p3g.org). In the relatively short history of regulating biotechnologies, it has become clear that the focus of law itself may change: for example, after the completion of the Human Genome Project, the notion of "genetic information" dominated almost the entire legal and ethical landscape. In contrast, this attitude is nowadays often regarded as "genetic exceptionalism". Furthermore, many other concerns have been crystallised in the context of tissue-based research and therapies, such as the scope of bodily integrity, privacy, freedom of research, the possibility and scope of enhancement. The recent developments of regenerative medicine, the presence of many international biobank consortia, and the new developments in synthetic biology indicate that the tissue issues have not yet reached the final complexity. In our book, we tackled systematic ethical and legal questions in the context of specific regions with the intention to analyse some key aspects of the current stage of development.

Index

MIX
Papier aus verantwortungsvollen Quellen
Paper from responsible sources
FSC® C105338

If you have any concerns about our products,
you can contact us on
ProductSafety@springernature.com

In case Publisher is established outside the EU,
the EU authorized representative is:
Springer Nature Customer Service Center GmbH
Europaplatz 3, 69115 Heidelberg, Germany

Printed by Libri Plureos GmbH
in Hamburg, Germany